Boundary Value Problems of Linear Partial Differential Equations for Engineers and Scientists

Boundary Value Problems of Linear Partial Differential Equations for Engineers and Scientists

Shien-siu Shu

World Scientific
Singapore • New Jersey • Hong Kong

Published by

World Scientific Publishing Co. Pte. Ltd.
P.O. Box 128, Farrer Road, Singapore 9128

U.S.A. office: World Scientific Publishing Co., Inc.
687 Hartwell Street, Teaneck NJ 07666, USA

Library of Congress Cataloging-in-Publication data is available.

BOUNDARY VALUE PROBLEMS OF LINEAR PARTIAL DIFFERENTIAL EQUATIONS FOR ENGINEERS AND SCIENTISTS

ISBN 9971-50-417-0
9971-50-418-9 (pbk)

Printed in Singapore by Fong and Sons Printers Pte. Ltd.

To My Family,

Wife Irene,

and Children: Alice, David, Frank, and Jeannie

PREFACE

A philosopher once said that mathematics is nothing but training; however, engineers have their own motivation to learn mathematics. Because they deal with complicated practical problems and are eager to find useful methods, engineers often face the task of constructing reasonable models, for which familiarity with analytical solutions of simple cases can throw light on the understanding of the more complex situations.

The present text is a revised version of the author's lecture notes in a graduate course of applied mathematics, developed at the Illinois Institute of Technology in the early fifties and expanded at Purdue University. The text is based on the idea that engineering students may find it more interesting to learn mathematics through the introduction of concrete examples. In carrying out this task, I have tried to organize the material in a logical order that transmits the package of mathematical knowledge and methods to the students in an efficient manner. Many problems, utilizing the existing laws of science, can be formulated in mathematical form; often the formulation leads to boundary-value problems in linear partial differential equations, for which solutions are then required. Thus, various standard methods of solution are naturally introduced, employing the techniques of Fourier series, orthogonal functions, Laplace and other transforms, Green's functions, Riemann's conformal mapping, etc. The only previous knowledge assumed as prerequisites is advanced calculus and elementary ordinary differential equations.

In order to enhance the students' enthusiasm for learning the material, exercises are chosen carefully. This enables the students not only to find their own way to solutions, but also strengthens their practical grasp of the subject under discussion in the text. Supplementary knowledge necessary for some treatments -- for instance, the elementary theory of analytic functions -- is included in an appendix. Cauchy's residue theorem is particularly important in the evaluation of definite integrals of the type that appear by employing operational methods.

The present text is not intended as a complete treatment of boundary-value problems in linear partial differential equations. A conscientious reader is encouraged to consult the additional material listed in the references.

Finally, the author wishes to thank Mrs. Terri Moore for her painstaking fine work in typing the notes in the present form.

TABLE OF CONTENTS

Boundary Value Problems of Linear Partial Differential Equations for Engineers and Scientists

CHAPTER I. INTRODUCTION

PARTIAL DIFFERENTIAL EQUATIONS IN MATHEMATICAL PHYSICS

A certain number of partial differential equations which occur frequently in classical mathematical physics are derived in this chapter. Emphasis is here given to the physical motivation of the boundary value problems.

The differential equations of this chapter are typical of those which are to be treated in this course. The boundary values, as they come out naturally from the actual problems, introduce the various standard methods of solution, employing the techniques of Fourier series and orthogonal functions, of Laplace transforms, of conformal mapping, of Green and Riemann, etc.

The partial differential equations which arise from the physical problems often represent the physical laws governing the motion or the process undergone. In order to predict the phenomena which are to occur at a later instant and at a definite place, some information or data must be known concerning the state of the physical system at a certain instant or at a certain place. These collections of data are referred to as the boundary values.

1. CONDUCTION OF HEAT IN HOMOGENEOUS AND ISOTROPIC SOLID

We shall first treat the simple case of the conduction of heat in an infinite slab or a wire with an insulated lateral surface. The formulation in the general case of the conduction of heat in a solid follows immediately by the similar arguments.

Let $u(x,t)$ be the temperature at a distance x from a fixed plane and at the time t.

The heat flux, F, in a given direction is defined as the amount of heat which flows in that given direction across a surface of unit area, in unit time. The equation of heat conduction is derived from the following laws:

(A) Newton's law* - the heat flux f in a given direction is proportional to the rate of change of the temperature in that direction.

(B) Heat balance - the net flux flowing into an element is entirely used to raise the temperature of the element. When the heat source is present, additional term to this effect should be taken into account.

For the present case, (A) gives

$$F = -ku_x(x,t), \quad k > 0$$

where: $u_x = \frac{\partial u}{\partial x}$

k = the conductivity (which may be variable such as a function of x).

Both k and u_x are assumed to be continuous up to their first derivatives. Consider now an element with volume $A\Delta x$. From (B), it is clear that

$$[-k(x)\, u_x(x,t) + k(x-\Delta x)\, u_x(x+\Delta x,t)]A = \frac{\partial}{\partial t}(uC\rho)A\Delta x$$

where: ρ = the density of the solid,

C = the specific heat.

Let the above equation be divided by $A\Delta x$ and let $\Delta x \rightarrow 0$. We have

$$\frac{\partial}{\partial x}\left(k\frac{\partial u}{\partial x}\right) = \frac{\partial}{\partial t}\rho C u \tag{1}$$

By similar arguments, the general equation for heat conduction in a solid is shown to be:

$$\frac{\partial}{\partial x}\left(k\frac{\partial u}{\partial x}\right) + \frac{\partial}{\partial y}\left(k\frac{\partial u}{\partial y} + \frac{\partial}{\partial z}\left(k\frac{\partial u}{\partial z}\right) = \frac{\partial}{\partial z}(\rho C u) \tag{2}$$

Where u(x,y,z,t) is assumed to be continuous up to its second derivatives in the space variables (x,y,z) and the time variable t, within the bounds of the solid.

---

*In the case of a non-isotropic solid, it is suggested that k should be replaced by a tensor and Newton's law may be written in the vector form $\bar{F} = -\bar{k}\cdot\nabla u$.

1.1 Special Cases.

(i) Physical Constants Independent of Temperature

k, ρ, and C are all of constant value. Then (2) becomes:

$$u_t = K\Delta u_{xx}$$

where

$$K = k/\rho C = \text{diffusivity}$$

$$\Delta = \frac{\partial^2}{\partial x^2} + \frac{\partial^2}{\partial y^2} + \frac{\partial^2}{\partial z^2}$$

(ii) The Steady Flow

If a further assumption is imposed that u is independent of t, then (3) becomes

$$\Delta u = 0 \tag{3a}$$

(iii) The One-dimensional Flow.

This simplest case of heat conduction in an infinite homogeneous, isotropic slab is still of practicl importance. The problem can be further simplified by making a suitable choice of units such that K = 1. This choice of units shall be assumed hereafter.

It is worthy of note that different physical interpretations can be given to the same equation for many other problems, such as, the diffusion of the concentration of a substance through a porous medium, an oscillating plane in the viscous flow and some special cases of the electric waves through transmission lines.

EXERCISES

(1) Derive the equation for heat conduction in an infinite circular bar by the same method as above with the assumption of cylindrical symmetry (and K = 1).

Answer: $r^{-1} \frac{\partial}{\partial r} (r\, u_r) = u_t$ where r is the distance from the axis of the bar.

(2) Assuming spherical symmetry, show that the equation for heat conduction in a sphere is

$$u_{rr} + \frac{2}{r} u_r = u_t \tag{4}$$

(3) Show that the equation (4) can be reduced to $v_{rr} = v_t$ if one takes the substitution $v = ru$.

1.2 The Boundary Values of an Infinite Slab

Let the slab be of width 1. It is intuitively clear that the temperature at any point and at an instant t is uniquely determined if the following conditions are prescribed:

(A) The initial temperature

$$u(x,0) = f(x) \ . \tag{5}$$

(B) The conditions at the ends $x = 0$ and $x = 1$, which shall be called the end control:

(i) <u>The prescribed temperatures at the ends.</u>

$$\begin{aligned} u(0,t) &= h(t) \\ u(1,t) &= g(t) \end{aligned} \tag{6a}$$

(ii) <u>The heat supply at the ends.</u>

$$\begin{aligned} u_x(0,t) &= h(t) \\ u_x(1,t) &= g(t) \end{aligned} \tag{6b}$$

(iii) <u>Radiation at the ends.</u> Let the medium surrounding the slab be kept at constant absolute temperature, say u_0. The radiation law of Stefan-Boltzmann states that the radiation of heat per unit area per unit time is proportional to $u^4 - u_0^4$, where u is the absolute temperature at one end.

It is evident that if the difference is small as compared with u_0, we have the approximate relations:

$$+ku_x(0,t) = a(u-u_0)$$

$$-ku_x(1,t) = a(u-u_0)$$

where a is constant, or

$$\begin{aligned} u_x(0,t) - hu(0,t) &= -hu_0 \\ u_x(1,t) + hu(1,t) &= hu_0 \end{aligned} \qquad (h = \frac{a}{k}) \tag{6c}$$

In short, the <u>boundary conditions for the slab</u> are of the following type:

$$u(x,0) = f(x)$$

$$L_1(u) = Au_x(1,t) + Bu(1,t) = g(t) \tag{5}$$

$$L_2(u) = Cu_x(0,t) + Du(0,t) = h(t)$$

where A, B, C, D are prescribed constants.

The reader is cautioned here to distinguish the conditions on the physical boundary (of the solid) and the boundary values in the xt-plane for the partial differential equation. The boundary values for the present problem are indicated in the following figure.

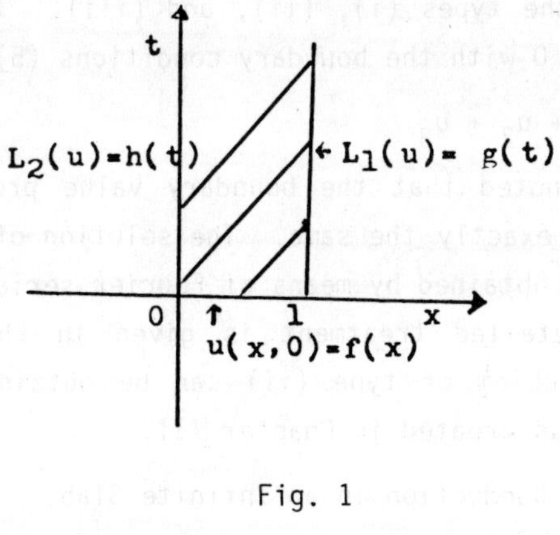

Fig. 1

We are interested in finding the solution of the partial differential equation (1) in the domain R satisfying the boundary conditions as indicated in the figure.

2. THE SUPERPOSITION PRINCIPLE

An operator L, such as $\frac{\partial^2}{\partial x^2} - \frac{\partial}{\partial t}$, is called homogeneous and linear if

$$L(u+v) = L(u) + L(v) \tag{7a}$$

$$L(au) = aL(v) \tag{7b}$$

where a is any constant.

If u_1, u_2, ... be such that $L(u_i) = 0$, $i = 1,2$ .., then any linear combination $u = \Sigma\, a_i\, u_i$ is a solution of $L(u) = 0$ provided the series can be operated on termwise by the operator L.

The boundary conditions (5) and (6) can be decomposed into 3 types:

(i) $g(t) \equiv h(t) \equiv 0$; $f(x)$ arbitrary

(ii) $f(x) \equiv h(t) \equiv 0$; $g(t)$ arbitrary

(iii) $f(x) \equiv g(t) \equiv 0$; $h(t)$ arbitrary

Let u_1, u_2, and u_3 be respectively the solution of the boundary value problems of the types (i), (ii), and (iii). The solution for the equation $L(u) = 0$ with the boundary conditions (5) and (6) is

$$u = u_1 + u_2 + u_3$$

It should be noted that the boundary value problems of types (ii) and (iii) are exactly the same. The solution of the problem of the type (i) can be obtained by means of Fourier series or orthogonal functions. The detailed treatment is given in Chapter II. The solution of the problem of type (ii) can be obtained by means of Laplace transforms as treated in Chapter III.

2.1 Example. Heat Conduction in an Infinite Slab.

The equation is $L(u) = u_{xx} - u_t = 0$. If the initial temperature at the ends $x = 0$ and $x = 1$ is kept at 0, we have the boundary values:

$$u(x,0) = f(x) \quad (\text{say} = 1) \tag{5'}$$

$$u(0,t) = 0 \tag{6'a}$$

$$u(1,t) = 0 \tag{6'b}$$

The procedure to solve this type of problem is:

(A) To find all solutions of the form

$$u = X(x)\, T(t) \tag{7}$$

which satisfy the <u>homogeneous</u> equation $L(u) = 0$, as well all the <u>homogeneous</u> conditions (6'a) and (6'b); and

(B) To satisfy the condition (5') by a linear combination of the solutions of (A), which are called eigenfunctions.

The method (A) is known as the <u>separation of variables</u>; (B) is known as the expansion of an arbitrary function in terms of eigenfunctions.

Substituting (7) in the equation $L(u) = 0$, we derive that

$$\frac{X''}{X} = \frac{T'}{T} = \lambda \text{ (a constant)} \tag{8}$$

where the prime (') denotes the differentiation. Since the quantities on the two sides of (8) are respectively functions of x and t, which are independent variables, they must both be a constant, λ.

Consider the equation

$$X'' - \lambda X = 0 \tag{9}$$

with the conditions

$$X(0) = 0 \tag{10a}$$

$$X(1) = 0 \tag{10b}$$

which are the consequences of (6'a) and (6'b).

A solution of (9) satisfying (10a) and (10b) other than the trivial one, $X = 0$, is called an eigenfunction. It shall be seen later that there is a discrete set of values of λ for which the eigenfunctions exist. Those values are called the eigenvalues.

We are going to show that the eigenvalues $\lambda < 0$. Suppose $\lambda > 0$. Let $\lambda = a^2$. Then:

$$X = Ae^{ax} + Be^{-ax} \quad .$$

By (10a) and (10b) we have:

$$A + B = 0$$

$$Ae^{a} + Be^{-a} = 0 \ , \quad \text{or} \quad A = B = 0 \ .$$

Thus $X \equiv 0$ and therefore X is not an eigenfunction. Now, set $\lambda = -a^2$. We have:

$$X = A \sin ax + B \cos ax$$

By (10a) and (10b), it follows that B = 0 and $a = n\pi$, where n is an integer. Thus all the distinct eigenfunctions are:

$$X_n = A \sin n\pi x \quad (n = 1, 2, 3 \ldots)$$

Consequently from (8), we have

$$T = A' \, e^{-n^2\pi^2 t}$$

Now, let us apply the superposition principle to find the solution of the problem by a formal procedure.

Put

$$u(x,t) = \sum_{n=1}^{\infty} A_n e^{-n^2\pi^2 t} \sin n\pi x \tag{11}$$

Where A_n are arbitrary constant to be determined by the condition (5'). At t = 0, we have:

$$f(x) = \sum_{n=1}^{\infty} A_n \sin n\pi x \quad . \tag{12}$$

Let us leave the question of possibility and validity of (12) at present. Assuming the identity in x, (12), we find a formal expresion for A_n in terms of the given function f(x). Multiplying (12) by sin m x for a fixed positive integer M and integrating from x = 0 to x = 1, we have the general formula for the (Fourier) coefficients:

$$A_m = 2 \int_0^1 f(x) \sin m\pi \, x \, dx \, , \tag{13}$$

The integration of the right hand side of (12) is performed term by term and the following fact of <u>orthogality</u> is utilized:

$$\int_0^1 \sin n\pi \, x \sin m\pi \, x \, dx = \begin{cases} 0 \, , & \text{for } m \neq n \\ \frac{1}{2} \, , & \text{for } m = n \, . \end{cases}$$

For the special case $f(x) \equiv 1$ for $0 > x > 1$, we have

$$A_m = \frac{2}{m\pi} (1 - (-1)^m) \quad \text{by} \quad (13) \, .$$

The solution of the heat conduction for that special initial temperature distribution is:

$$u(x,t) = \frac{2}{\pi} \sum_{n=1}^{\infty} \frac{1}{n} \{1 - (-1)^n e^{-n^2\pi^2 t}\} \sin n\pi x$$

$$= \frac{4}{\pi} \; (e^{\pi 2t} \; \sin\pi \; x + \frac{1}{3} e^{-9\pi^2 t} \sin 3\pi x$$

$$+ \frac{1}{5} e^{-25\pi^2 t} \sin 5\pi x + \ldots)$$

3. FOURIER EXPANSIONS

For a given integrable function f(x) in [0,1], the Fourier coefficients are defined by (13). The series:

$$\sum_{n=1}^{\infty} A_n \sin n\pi x$$

in (12) is defined as the Fourier sine series of f(x) on [0,1].

The subject of Fourier series has been one of the much developed branches of analysis in modern mathematics. One of the important questions is that under what condition the Fourier series converges? We merely mention a result which is useful in many practical problems. The conditions can be very much weakened.

Let us consider, for simplicity, only those functions f(x)* such that the following Dirichlet conditions are satisfied:

(A) f(x) is continuous except a finite number of finite jumps in [-1,1].

(B) f(x) has a finite number of maxima and minima in [1,1].

For any point x = a in [-1,1], the limiting values of f(x) exist as x tends to a from the left hand side of x = a and from the right hand side of x = a.

We shall denote them by:

---

*The conditions (A) and (B) are special cases of a conveniently stated condition that f(x) is one of "bounded variation." As a matter of fact, the study of Fourier series leads to the consideration of a much broader class of functions. If a student is interested in this direction, a standard text on the subject should be consulted. For instance, the last chapter of Titchmarsh's Theory of Functions.

$$f(a-0) = \lim_{\varepsilon\to 0} f(a-\varepsilon) \quad \text{or} \quad f(a-) ,$$

and

$$f(a+0) = \lim_{\varepsilon\to 0} f(a+\varepsilon) \quad \text{or} \quad f(a+) .$$

If the function f(x) is continuous at x = a, it is clear that:

$$\frac{1}{2}[f(a-0) + f(a+0)] = f(a) .$$

We merely state the following theorem which answers the question concerning the expansion of an arbitrary function in terms of sine and cosine functions. The theorem in a somewhat more general form concerning the expansion in terms of eigenfunctions is stated in Chapter II. (cf. the proof in the appendix of Chapter III.)

Theorem. <u>If f(x) satisfies the assumptions (A) and (B), then the Fourier series:</u>

$$\frac{B_0}{2} + \sum_{n=1}^{\infty} (A_n \sin n\pi x + B_n \cos n\pi x)$$

where:

$$A_n = \int_{-1}^{1} f(x) \sin n\pi x \, dx$$

$$B_n = \int_{-1}^{1} f(x) \cos n\pi x \, dx \qquad (n = 0, 1, 2,)$$

<u>converges to the value:</u>

$$\frac{f(x+0) + f(x-0)}{2}$$

Indeed, the convergence is uniform in x if f(x) is continuous and f(-1) = f(1). The limiting values of f(x) at the end points x = -1 and x = 1 are obtained by defining the function f(x) with the period 2: f(x) = f(x+2).

EXERCISES

(1) Show that if g(x) satisfies the condition (A) and (B) in the interval [-L,L] instead of [-1,1], then:

$$\frac{g(x-0) + g(x+0)}{2} = \frac{B_o}{2} + \sum_{n=1}^{\infty} [A_n \sin \frac{n\pi}{L} x \; B_n \cos \frac{n\pi}{L} x]$$

where:

$$A_n = \frac{1}{L}\int_{-L}^{L} g(x) \sin \frac{n\pi}{L} x \, dx$$

$$B_n = \frac{1}{L}\int_{-L}^{L} g(x) \cos \frac{n\pi}{L} x \, dx$$

$$n = 0, 1, 2, \ldots$$

HINT: Set $x = L\xi$ and $g(x) = g(L\xi) = f(\xi)$

(2) If, in addition to the exercise (1), $g(x) = -g(-x)$, then the Fourier series of $g(x)$ has only sine terms and:

$$A_n = \frac{2}{L}\int_{0}^{L} g(x) \sin \frac{n\pi}{L} x \, dx$$

State the similar theorem for the case $g(x) = g(-x)$. Every function defined in $[0,L]$ can be either considered as an odd function or an even function in $[-L,L]$.

(3) Take $f(x) = x^2$ in interval $[0,1]$. Show by the Fourier Cosine Expansion theorem, that

$$1 + \frac{1}{2^2} + \frac{1}{3^2} + \ldots. + \frac{1}{n^2} = \frac{\pi^2}{6}$$

3.1 Convergence of the Fourier Series

In order to show that the solution obtained formally in Section 2 satisfies the equation $L(u) = 0$, it is sufficient to prove that the series (12) can be differentiated term by term with respect to either independent variable for $t > 0$ and $0 < x < 1$. This is established if the differentiated series is uniformly convergent for $t \geqq t_o > 0$ and $0 \geqq x \leqq 1$. To prove this, the following lemma due to Abel is often useful:

Abel's lemma. Let ε_n be a positive, monotonic decreasing sequence. Let a_n be any sequence such that $S_m = a_1 + \ldots + a_m$ is always bounded, i.e.,

$$A \leqq S_m \leqq B \quad \text{for} \quad m = 1, 2, 3 \ldots$$

where A and B are two fixed numbers, then:

$$\varepsilon_1 A \leqq \sum_{n=1} \varepsilon_n a_n \leqq \varepsilon_1 B .$$

Proof: Put $a_n = S_n - S_{n-1}$

$$K = \sum_{n=1}^{m} \varepsilon_n a_n = \varepsilon_1 a_1 + \sum_{n=2}^{m} \varepsilon_n (S_n - S_{n-1})$$

$$= \sum_{n=1}^{m} \varepsilon_n S_n - \sum_{n=2}^{m} \varepsilon_n S_{n-1}$$

$$= \varepsilon_1 S_1 + \varepsilon_2 S_2 + \varepsilon_3 S_3 + \dots \varepsilon_m S_m$$

$$- (\varepsilon_2 S_1 + \varepsilon_3 S_2 + \dots + \varepsilon_m S_{m-1})$$

$$= (\varepsilon_1 - \varepsilon_2) S_1 + (\varepsilon_2 - \varepsilon_3) S_2 + \dots + (\varepsilon_{m-1} - \varepsilon_m) S_{m-1} + \varepsilon_m S_m$$

Since: $-\varepsilon_n + \varepsilon_{n-1} > 0$, we have:

$$K \geqq (\varepsilon_1 - \varepsilon_2) A + (\varepsilon_2 - \varepsilon_3) A + \dots + (\varepsilon_{m-1} - \varepsilon_m) A + \varepsilon_m A = \varepsilon_1 A$$

Similarly,

$$K \leqq \varepsilon_1 B .$$

Cor. If:

$\sum_{n=1}^{m} U_n(x)$ is uniformly bounded; i.e.,

$$\left| \sum_{n=1}^{m} U_n(x) \right| \leqq A$$

for m = 1, 2, 3 ... and $a \leqq x \leqq b$, where A is a definite constant, and if $\varepsilon_n - \varepsilon_{n+1} \geqq 0$ for n sufficiently large and $\lim_{n\to\infty} \varepsilon_n = 0$, then $\sum_{n=1}^{\infty} \varepsilon_n U_n(x)$ converges uniformly in [a,b].

The proof is left to the student.

3.2 The Precise Formulation of the Boundary Value Problem

As we shall later see, in order to assure the uniqueness of the solution, some continuity conditions have to be imposed on the solution. Then the solution obtained by the formal procedure is shown to satisfy these continuity conditions. For instance, the boundary value problem of the heat conduction in the slab in 2.1 can be stated precisely as follows:

$$L(u) = 0 \qquad (t > 0, \quad 0 < x < 1)$$

$$u(x,+0) = \frac{f(x+0) + f(x-0)}{2} \qquad (0 < x < 1) \tag{5'}$$

$$u(+0,t) = 0 \qquad (t > 0) \tag{6a'}$$

$$u(1-0,t) = 0 \qquad (t > 0) \tag{6b'}$$

Besides, $u(x,t)$ is required to be uniformly bounded.

Now we are ready to show <u>the existence</u> of the solution of the boundary value problem as formulated above for the case $f(x)$ is continuous in $[0,1]$ and $f(0) = f(1) = 0$. The general case when $f(x)$ satisfies the conditions (A) and (B) in Section 3 can be proved by a similar argument with slight modifications (cf. Exercise (2) following this section).

We are thus going to prove that the solution (12) actually satisfies all the above conditions.

(i) Since the series $\sum_{n=1}^{\infty} A_n \sin n\pi x$ converges uniformly by the Fourier expansion theorem in Section 3, the series $\sum_{n=1}^{\infty} e^{-n^2\pi^2 t} (A_n \sin n\pi x)$ is certainly bounded uniformly for $t \geqq 0$ and $0 \geqq x \quad 1$ by Abel's lemma.

(ii) The differentiated series (term by term) of (12) with respect to either one of the independent variables to any order is of the form:

$$\sum_{n=1}^{\infty} a_n c A_n \sin n x \text{ , where } a_n = n^a e^{-n^2\pi^2 t} \tag{14}$$

and a, c are absolute constants independent of n. Now a_n is monotonically decreasing and positive for n large enough and

$$\sum_{n=1}^{\infty} cA_n \sin n\pi x$$

is uniformly convergent. Thus by the corollary of Abel's lemma, the series (14) coverges uniformly for $0 \leqq x \leqq 1$ and $t \geqq t_0 \geqq 0$. Hence the (term by term) differentiation is justified by a theorem in the advanced calculus that a series can be differentiated term by term if the differentiated series is uniformly convergent. Therefore:

$$L(u) = \sum_{n=1}^{\infty} A_n L(U_n) = 0 .$$

(iii) Since a uniformly convergent series of continuous functions is continuous, the conditions (6a') and (6b') are evident by (i). So is the condition (5'). Thus the <u>existence</u> is established.

<u>EXERCISES</u>

(1) A series $\sum_{n=1}^{\infty} U_n(x)$ is called boundedly convergent in [a,b] if there is a constant A such that:

$$\left|\sum_{n=1} U_n(x)\right| < A , \qquad \begin{array}{l} a \leqq x \leqq b \\ m = 1, 2, 3, \ldots \end{array}$$

and it converges everywhere in [a,b]. Show that the Fourier series of

$$\sigma(x) = \begin{cases} \dfrac{1-x}{2} & \text{for } 0 \leqq x \leqq 1 \\ \dfrac{-1-x}{2} & \text{for } -1 \leqq x \leqq 0 \end{cases}$$

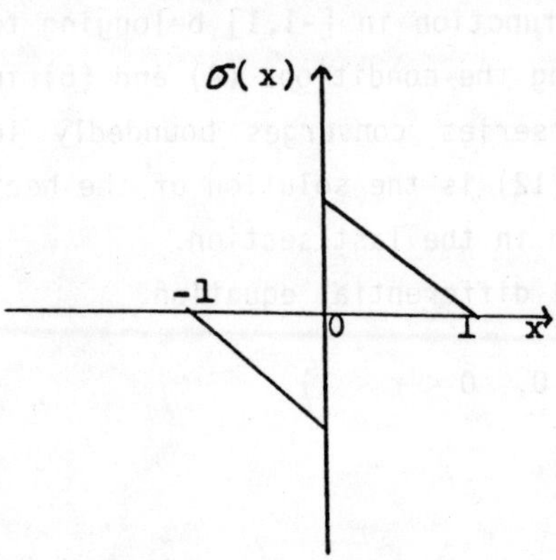

Fig. 2

is $\frac{2}{\pi}\sum_{n=1}^{\infty}\frac{\sin n\pi x}{n}$. The series converges to $\frac{\sigma(+0)+\sigma(-0)}{2}=0$ at $x = 0$. Define $\sigma(0) = 0$ and $\sigma(x+1) = \sigma(x)$. Show that Fourier series is boundedly convergent in $[-1,1]$.

HINT: Show that (a) $S_m(x) = \sum_{n=1}^{m} \cos n\pi x$

$$= \tfrac{1}{2}\left(-1 + \frac{\sin (m+\frac{1}{2})\pi x}{\sin \frac{1}{2}\pi x}\right)$$

(b) $$\sum_{n=1}^{m}\frac{\sin n\pi x}{n} = \int_0^x \pi S_m(\xi)\, d\xi$$

(c) $$\left|\int_0^x \left\{\frac{1}{\sin \frac{1}{2}\pi\xi} - \frac{2}{\pi\xi}\right\}\sin (m+\tfrac{1}{2})\pi\xi \; d\xi\right| < B$$

$(-1 \leqq x \leqq 1)$ where B is a constant independent of m.

(d) $$\frac{2}{\pi}\int_0^x \frac{\sin (m+\frac{1}{2})\pi\xi}{\xi}\,d\xi = \frac{2}{\pi}\int_0^{(m+\frac{1}{2})\pi x}\frac{\sin y}{y}\,dy$$

by putting $(m+\frac{1}{2})\pi\xi = y$.

(e) The function $f(x) = \frac{2}{\pi}\int_0^{(m+\frac{1}{2})\pi X}\frac{\sin y}{y}\,dy$ is bounded for all values of X since $|f(A_1) - f(A_2)|$ can be made as small as one pleases if both A_1 and A_2 are large enough.

(2) Assume that if f(x) is an odd function in [-1,1] belonging to the class of functions satisfying the conditions (A) and (B) in section 3, then its Fourier series converges boundedly in [-1,1]. Show that the series (12) is the solution of the heat conduction problem as formulated in the last section.

(3) Find the solution of the partial differential equation.

$$U_t = U_{rr} + \frac{2}{r} U_r \qquad (t > 0, \quad 0 < r < 1)$$

with the boundary conditions:

$$U(r,+0) = f(r)$$

$$U(1-0,t) + U_r(1-0,t) = 0$$

and U(r,t) is bounded.

HINT. Set u = rU(r,t). Find the boundary conditions for v and reduce the problem to that of heat conduction in a slab.

(4) Find the solution of the partial differential equation.

$$U_{xx} + U_{yy} = x^2 + 2y^2 + xy^3 \quad (x^2 + y^2 < 1)$$

with the boundary condition (in xy-plane)

$$U = 0 \quad \text{on} \quad x^2 + y^2 = 1$$

HINT. Find a particular solution of the partial differential equation by any method:

$$U_0(x,y) = \frac{x^4}{12} + \frac{y^4}{6} + \frac{xy^5}{20}$$

Use the polar coordinates and apply the transformation of the dependent variable

$$v(r,\Theta) = U(r \cos \Theta, r \sin \Theta) - U_0(r \cos \Theta, r \sin \Theta)$$

Show that:

$$v_{rr} + \frac{1}{r} v_r + \frac{1}{r^2} v_{\Theta\Theta} = 0 \ .$$

Set the boundary condition for v on r = 1 and solve the problem by separation of variables.

(5) Solve the following boundary-value problem:

$$U_{rr} + \frac{1}{r} U_r + \frac{1}{r} U_{\Theta\Theta} = 0$$

where:

$$U(1-0,\Theta) = f(\Theta) = \sum_{n=0}^{\infty} (A_n \sin n\Theta + B_n \cos n\Theta) .$$

A_n and B_n are given constants. The function U is assumed to be bounded and periodic in Θ of period 2π.

(6) <u>Improvement of convergence of some Fourier series.</u>

We have already indicated in (1) a general method. When the general Fourier coefficient A_n has the asmptotie expansion (for large values of n):

$$A_n = \frac{\alpha_1}{n} + \frac{\alpha_2}{n(n+1)} + \ldots + \frac{\alpha_r}{n(n+1)(n+r-1)} + 0\left(\frac{1}{n^{r+1}}\right)$$

then it can be separated as (r+1) series, of which the first r series can be summed up while the (r+1) th converges as one with coefficient $0(\frac{1}{n^{r+1}})$.

For example:

$$S = \sum_{n=1}^{\infty} \frac{n^2+1}{n^4+2n^2} \sin n\pi x$$

$$\frac{n^2+1}{n^4+2n^2} = \frac{1}{n(n+1)} + \frac{n^2 - n+1}{n^2(n^2+2)(n+1)}$$

$$= \frac{1}{n(n+1)} + \frac{1}{n(n+1)(n+2)} + R_n$$

where

$$R_n = \frac{(n-1)(n-2)}{n^2(n^2+2)(n+1)(n+2)} = 0\left(\frac{1}{n^4}\right)$$

Now, we have:

$$S = S_2 + S_3 + S^*$$

where

$$S_2 = \sum_1^{\infty} \frac{\sin n x}{n(n+1)}$$

$$S_3 = \sum_1^{\infty} \frac{\sin n x}{n(n+1)(n+2)}$$

$$S^* = \sum_1^{\infty} \bar{R}_n \sin nx$$

Find the answer of S_1 and S_2 by considering the following identities:

$$z = e^{x\pi i}$$

$$P_r = \sum_{n=1}^{\infty} \frac{z^n}{n(n+1)(n+r-1)} = \frac{1}{z^{r-1}} \sum_{n=1}^{\infty} \frac{z^{n+r-1}}{n...(n+r-1)}$$

$$= \frac{-1}{z^{r-1}(r-2)!} \int_0^z \log(1-\xi)\ (z-\xi)^{r-2} d\xi$$

$$= \frac{1}{(r-2)!} \left[\frac{1}{(r-1)z^{r-1}} \int_0^z (z-\xi)^{r-1} \frac{d\xi}{1-\xi}\right] \quad \text{for } r \geq 2.$$

Now $S_r = \text{Im } P_r(z)$. Observe that S^* converges rather fast.

4. THE VIBRATING STRING

The transverse displacements of a stretched string satisfy an important differential equation called the wave equation. The equation governs the propagation of small disturbances in a medium. The case of a vibrating sting is the simplest actual example which can be visualized easily.

The following assumptions are imposed on the problem:

(A) The string is perfectly flexible; i.e., the tension is always acting along the string.

(B) The weight of the string per unit length is so small as compared with the tension that the gravitational force can be neglected.

(C) The displacement is so small as compared with the total length of the string so that the slope everywhere is small. The smallness is equivalent to assuming that its square can be neglected. Such a simplification is called "linearization." Because of this assumption, the tension is considered as a constant along the string.

Let the string at the equilibrium position be on the x-axis of a rectangular coordinate system. Let U = U(x,t) be the vertical displacement of the point x on the string at instant t. Let T be the tension of the string.

Consider the element of the string between x and x + x. By (A) and (B), the vertical component of the force acting on the element is:

$$\left.\frac{TU_x}{(1+U_x^2)^{\frac{1}{2}}}\right|_{x+\Delta x} - \left.\frac{TU_x}{\sqrt{1+U_x^2}}\right|_x = \frac{\partial}{\partial x}\left(T\frac{U_x}{\sqrt{1+U_x^2}}\right) \quad \Delta x+\varepsilon\Delta x$$

where $\varepsilon\to 0$ as $x\to 0$, if the derivatives are continuous. On the other hand, by Newton's law of the motion this is equal to:

$$\rho\Delta x \cdot \frac{\partial^2 U}{\partial t^2} \quad .$$

where ρ is the mass of the string per unit length. Dividing by x and making $\Delta x\to 0$, we have:

$$\frac{\partial}{\partial x}\left(T\frac{U_x}{(1+U_x^2)^{\frac{1}{2}}}\right) = \frac{\partial^2 U}{\partial t^2} \tag{14}$$

By (C), the equation (14) is simplified as

$$U_{xx} - \frac{1}{a^2} \quad U_{tt} = 0$$

where $a^2 = \frac{T}{\rho}$. The constant a, as we shall see immediately*, is the speed of propagation of the small disturbances.

EXERCISES

(1) Derive the equation for the vibrating membrane with the same assumptions as made in Section 4.

---

*Compare Ex. 5

Ans. $U_{xx} + U_{yy} - \frac{1}{a^2} U_{tt} = 0$

where:

$$a = \frac{T}{\rho} = \frac{\text{surface tension per unit length}}{\text{density of the membrane for unit area.}}$$

(2) Find the solution of a vibrating string of length 2 with the end points fixed while it is plucked at the middle when at rest, and is then released.

HINT: In addition to the equation (15), the boundary conditions are

$$U(x,0) = \begin{cases} h\,x & (0 \leq x \leq 1) \\ h(2-x) & (1 \leq x \leq 2) \end{cases}$$

$$U_t(x,0) = 0$$

$$U(0,t) = 0$$

$$U(2,t) = 0$$

Solve the equation (15) by the method of separation of variables and Fourier series similar to that used in the heat conduction problem.

Ans. $U(x,t) = \frac{4h}{\pi^2} \sum_{n=1}^{\infty} \frac{\sin \frac{n}{2}\pi}{n^2} \{\sin \frac{n\pi}{2}(x+at) + \sin \frac{n\pi}{2}(x-at)\}$

(3) Show that if a damping force, say, proportional to the velocity, such as the air resistence, acts upon the membrane of Exercise (1), the equation of the motion is:

$$U_{xx} + U_{yy} = \frac{1}{a^2} U_{tt} + bU_t$$

where b is a positive constant.

(4) By assuming Hooke's law, show that the equation for a longitudinal vibration of an elastic bar with the cross section A(x) and Young's modulus E is:

$$E \frac{\partial}{\partial x}(A(x)U_x) = \rho A(x)\, U_{tt}$$

where $U = U(x,t)$ = the longitudinal displacement, ρ = the density of the bar per unit volume.

(5) Show that the general solution of the equation:

$$U_{xx} - \frac{1}{a^2} U_{tt} = 0 \text{ is } U(x,t) = F(x-at) + G(x+at), \text{ where:}$$

F and G are arbitrary functions continuous up to their second derivatives, except at a finite number of values in a finite interval. F(x-at) represent a wave travelling to the right with the velocity a, and G(x+at) represents a similar wave travelling to the left.

HINT: Show that the wave equation reduces to:

$$U_{\xi\eta} = 0$$

where:

$$\xi = x-at .$$

$$\eta = x+at .$$

5. TRANSMISSION LINES

Let v = the voltage = v(x,t)

i = the current = i(x,t)

R = the resistence per unit length of the line

L = the inductace per unit length

C = the capacitace per unit length

G = the leakage conductace per unit length.

Consider an element between x and x+Δx of a transmission line. The equations for such a line are obtained from the following facts:

(A) Ohm's law. $E = R^* i$

where:

$$E = E_1 + E_2$$

$$E_1 = \text{applied E.M.F.}$$

$$E_2 = \text{induced E.M.F.}$$

(B) Faraday's law.

(B_1) (the magnetic flux F induced by the current i) = $L^* i$

(B_2) induced E.M.F. = $-\frac{\partial F}{\partial t}$

or $E_2 = -L^* \frac{\partial i}{\partial t}$

(C) The charges entering the element per unit time are partly to build up the potential and partly to compensate for the leakage conductance due to dielectric loss. The former is equal to:

$$\frac{\partial C^* v}{\partial t}$$

and the latter is equal to Gv.

From (A) and (B), we have:

$$Ri\Delta x = v(x,t) - v(x+\Delta x,t) - L \frac{\partial i}{\partial t}\Delta x .$$

Divide by Δx and let $\Delta x \to 0$. It follows that:

$$v_x + Li_t + Ri = 0$$

From (C),

$$i(x,t) - i(x+\Delta x,t) = [C \frac{\partial v}{\partial t} + Gv] \Delta x .$$

Thus, we have:

$$Cv_t + Gv + i_x = 0 . \tag{17}$$

The elimination of one of the dependent variables is done by differentiation. The algebra involved would be exactly the same by considering the differential operators as if quantities:

$$D = \frac{\partial}{\partial t} \quad \text{and} \quad \delta = \frac{\partial}{\partial t} .$$

By Cramer's rule applied to a system of linear equations, we have:

$$\begin{vmatrix} D, & L\delta + R \\ C\delta + G, & D \end{vmatrix} \quad v = [D^2 - CL\delta^2 - (CR+LG)\delta - RG)] \quad v = 0$$

or

$$v_{xx} - CLv_{tt} - (CR+LG)v_t - RGv = 0 \tag{18}$$

It is seen readily that the current i satisfies the same equation as (18).

EXERCISES

(1) Flow of Perfect Fluids

Let the physical quantities such as the pressure p, the density ρ and the velocity components u, v and w along the rectangular axes be observed at a point (x,y,z) of the space. Thus they are functions of the space variables and time t. Following a particle of the fluid, the x, y and z coordinates of the particle are functions of t. The following notation is used:

$$\frac{d}{dt} = u\frac{\partial}{\partial x} + v\frac{\partial}{\partial y} + w\frac{\partial}{\partial z} + \frac{\partial}{\partial t} .$$

Derive the following equations:

(A) The equations of motion.

$$\rho \frac{du}{dt} = -\frac{\partial p}{\partial x} \tag{19}$$

$$\rho \frac{dv}{dt} = -\frac{\partial p}{\partial y} \tag{20}$$

$$\rho \frac{dw}{dt} = -\frac{\partial p}{\partial z} \tag{21}$$

(B) The equation of continuity of mass

$$\rho \left(\frac{\partial u}{\partial x} + \frac{\partial v}{\partial y} + \frac{\partial w}{\partial z}\right) + \frac{d\rho}{dt} = 0 \tag{22}$$

HINT: The physical facts which are utilized to derive the equations are:

(A) The pressure acts always normally to a surface of fluid. In applying Newton's second law, the weight is neglected as compared with the surface force due to the pressure of the fluid. Viscosity is also neglected.

(B) The mass which enters into an element in space per unit time is exactly equal to the amount required to increase the density of the fluid in the element per unit time.

(2) If the fluid is isentropic, it is derived in thermodynamics of a perfect fluid that p/ρ^{γ} = constant throughout the fluid,

where $\gamma = C_p/C_v$

$$= \frac{\text{specific heat at constant pressure}}{\text{specific heat at constant volume}}$$

Suppose that a small disturbance is given to the fluid at rest. Derive the equation of sound.

HINT: Let $p = p_o + \bar{p}$, $= \rho_o + \bar{\rho}$, $u = \bar{u}$, $v = \bar{v}$, $w = \bar{w}$, where p_o and ρ_o values at rest. Substitute the above four expressions and the "p-ρ" relation into the equations derived in (1), and show, by neglecting the terms of degree higher than the first, that both the quantities p and ρ satisfy the wave equation:

$$\Box\varphi = 0 ,$$

where

$$\Box = \frac{\partial^2}{\partial x^2} + \frac{\partial^2}{\partial y^2} + \frac{\partial^2}{\partial z^2} - \frac{1}{a^2}\frac{\partial^2}{\partial t^2}$$

and: $a^2 = \frac{\gamma p_o}{\rho_o}$. The equation is also satisfied by φ, the velocity potential, which is so defined that $\varphi_x = \bar{u}$, $\varphi_y = \bar{v}$, $\phi_2 = \bar{w}$.

6. REDUCTION OF A GENERAL LINEAR PARTIAL DIFFERENTIAL EQUATION OF SECOND ORDER TO ITS CANONICAL FORM

The equations for the transmission lines in Section 5 are of the "hyperbolic" or "parabolic" type. We are going to show that a general equation of second order:

$$Au_{xx} + 2Bu_{xy} + Cu_{yy} + Du_x + Eu_y + Fu = 0 \tag{23}$$

where the coefficients are constants, can be reduced by the change of both the dependent variable and the independent variables, to one of the following canonical forms:

(i) Hyperbolic Type ($\delta > 0$)

$$v_{\xi\xi} - v_{\eta\eta} + Kv = 0$$

(ii) Elliptic Type ($\delta < 0$)

$$v_{\xi\xi} + v_{\eta\eta} + Kv = 0$$

(iii) Parabolic Type ($\delta = 0$)

$$v_{\xi\xi} - v_{\eta} = 0 \quad \text{or} \quad v_{\xi\xi} = Kv = 0$$

where: $\delta = B^2 - AC$, and K is a constant.

Let us first apply the linear transformation:

$$\xi = ax + bu$$
$$\eta = cx + dy \tag{24}$$

where the determinant:

$$\begin{vmatrix} a, & b \\ c, & d \end{vmatrix} \neq 0 .$$

By this transformation, (23) is transformed into the following equation of a similar form:

$$A'u_{\xi\xi} + 2B'u_{\xi\eta} + C'u_{\eta\eta} + D'u_{\xi} + E'u_{\eta} + F'u = 0 \tag{25}$$

where:

$$\begin{aligned} A' &= Aa^2 + 2B\,ab + Cb^2 \\ B' &= Aac + B\,(ad+bc) + C\,bd \\ C' &= Ac^2 + 2B\,cd + Cd^2 \end{aligned} \tag{26}$$

The algebra here is exactly the same as that of the transformation of a quadratic form. For instance, it is a simple matter to verify the following identity by the multiplication of determinants (row by column) and by means of (26):

Theorem
$$\begin{vmatrix} A', & B' \\ B', & C' \end{vmatrix} = \begin{vmatrix} a, & b \\ c, & d \end{vmatrix} \cdot \begin{vmatrix} A, & B \\ B, & C \end{vmatrix} \cdot \begin{vmatrix} a, & c \\ b, & d \end{vmatrix} \tag{27}$$

Therefore the sign of $\delta = B^2 - AC$ is invariant under linear transformations (24).

Case (i) $\delta > 0$. So is $\delta' = B'^2 - A'c' > 0$

Let us determine the coefficients of the transformation (24) such that:

$$A' = C' = 0 .$$

Then $B'^2 > 0$. Suppose $A \neq 0$. Let $\frac{a}{b} = \lambda_1$ and $\frac{c}{d} = \lambda_2$. From (24) and (26), it follows that λ_1 and λ_2 are the roots of the equation

$$A\lambda^2 + 2B\lambda + C = 0 \tag{28}$$

Thus the special transformation:

$$\begin{aligned} \xi_1 &= \lambda_1 x + y \\ \eta_1 &= \lambda_2 x + y \end{aligned} \tag{29}$$

where:

$$\begin{vmatrix} \lambda_1, & 1 \\ \lambda_2, & 1 \end{vmatrix} = \lambda_1 - \lambda_2 = \frac{\pm 2\delta^{\frac{1}{2}}}{A} \neq 0$$

transforms the given partial differential equation to:

$$u_{\xi_1\eta_1} + u_{\xi_1} + \beta u_{\xi_1} + \gamma u = 0 \tag{30}$$

where α, β and γ are constants.

We shall see later (Chapter VI) the straight lines ξ_1 = constant and η_1 = constant are those curves across which the discontinuity of the derivatives of the u may occur. (In the case of the vibrating string, we know that $\xi_1 = x - at$ and $\eta_1 = x+at$ are curves on the xt plane along which discontinuities propagate.) These curves are called characteristics. We may interpret the reduction of the equation to (30) by the transformation (29) as the simplification rendered by taking the characteristics as the coordinate axes. This point of view gives a general criterion in carrying out the transformations for other types of partial differential equations.

Next, let $u = e^{-(\beta\xi_1+\alpha\eta_1)}v$ (31)

Then (30) is further reduced to:

$$v_{\xi_1\eta_1} + \gamma v = 0 \tag{32}$$

Finally, by the following transformation,

$$\xi_1 = \frac{1}{2}(\xi+\eta)$$

$$\eta_1 = \frac{1}{2}(\xi-\eta)$$

(32) reduces to (i).

Case (ii) $\delta < 0$. $(AC > 0)$

The analogue of the transformations (29) and (33) is:

$$\begin{aligned} \xi &= \xi_1 + \eta_1 = (\lambda_1+\lambda_2)\,x + 2y = -\frac{2B}{A}x + 2y \\ \eta &= \frac{1}{i}(\xi_1-\eta_1) = \frac{\lambda_1-\lambda_2}{i}\,x = \frac{+2\sqrt{-\delta}}{A}\,x \end{aligned} \tag{34}$$

where λ_1 and $\lambda_2 = \bar{\lambda}_1$ are roots of the equation (28); i.e.,

$$\lambda_1 + \lambda_2 = -\frac{2B}{A}$$

$$\frac{1}{i}(\lambda_1-\lambda_2) = \frac{\pm 2\sqrt{-\delta}}{A}$$

Observe that the transformation (34) is real. Then the partial differential equation is reduced to:

$$u_{\xi\xi} + u_{\eta\eta} + \alpha u_\xi + \beta u_\eta + \gamma_u = 0 \tag{35}$$

By setting

$$u = e^{-\frac{1}{2}(\alpha\eta+\beta\eta)}\,v \tag{36}$$

the equation (35) reduces to (ii).

Case (iii) $\delta = 0$

Suppose that $A \neq 0$. The terms of the second derivatives in the original partial differential equation can be written in the form:

$$A\left(\frac{\partial}{\partial x}+\frac{B}{A}\frac{\partial}{\partial y}\right)^2 u .$$

Thus the transformation:

$$x = \xi$$
$$y = \frac{B}{A}\ \xi + \eta \tag{37}$$

or:

$$\xi = x$$
$$\eta = \frac{-B}{A}\ x + y \tag{38}$$

carries the equation into:

$$u_{\xi\xi} + \alpha u_{\xi} + \beta u_{\eta} + \gamma u = 0 \tag{39}$$

Now we have to distinguish between the cases $\beta \neq 0$ and $\beta = 0$. For $\beta \neq 0$, set:

$$u = e^{-\frac{1}{2}\alpha\xi + k\eta}v$$

where $k = \frac{\alpha^2 - \beta\alpha\gamma}{\beta}$ (39) becomes $v_{\xi\xi} - \beta v_{\eta} = 0$. For $\beta = 0$, the substitution:

$$u = e^{-\frac{1}{2}\ x\xi}v$$

carries (39) to:

$$v_{\xi\xi} - Kv = 0.$$

EXERCISES

(1) Show that if (1) $\xi = f(x,y)$ is a solution of the partial differential equation (2) $A(x,y)\xi_x + B(x,y)\xi_y = 0$, then for a constant value ξ, (1) is the solution of the ordinary differential equation:

$$\frac{dx}{A(x,y)} = \frac{dy}{B(x,y)} \tag{3}$$

Prove also the converse.

(2) Show that if the coefficients A, ---, and F are functions of x and y, the transformation:

$$\xi = \xi(x,y)$$
$$\eta = \eta(x,y) \tag{40}$$

where:

$$\begin{vmatrix} \xi_x, & \xi_y \\ \eta_x, & \eta_y \end{vmatrix} \neq 0 ,$$

carries the equation to the form (25) where:

$$A' = Q(\xi) = A\xi x^2 + 2B\xi_x\xi_y + C\xi_y^2$$

$$B' = Q(\xi,\eta) = A\xi_x\eta_x + B(\xi_x\eta_y + \xi_y\eta_x) + C\xi_y\eta_y$$

$$C' = Q(\eta) .$$

Also show that:

$$-\delta' = \begin{vmatrix} A', & B' \\ B', & C' \end{vmatrix} = -\delta j^2$$

(3) Prove that there exists a real transformation (40) which carries the partial differential equation to one of the following forms:

(i) $u_{\xi\xi} - u_{\eta\eta} + D'(\xi,\eta)u_\xi + E'(\xi,\eta)u_\eta + F'(\xi,\eta)\,u = 0$ for $\delta > 0$.

(ii) $u_{\xi\xi} + u_{\eta\eta} + D'u_\xi + E'u_\eta + F'u = 0$, for $\delta < 0$.

(iii) $u_{\xi\xi} + D'u_\xi + E'u_\eta + F'u = 0$, for $\delta = 0$.

(4) Transform the following equation:

$$u_{xx} + yu_{yy} = 0,$$

into the canonical form according to $y > 0$ or $y < 0$.

7. EQUATIONS OF HIGHER ORDER. TRANSVERSE VIBRATIONS OF AN ELASTIC BEAM

Partial differential equations of higher order occur very often in mathematical physics. For a certain type of boundary type problem, the order and the type of the equation do not play any role concerning the method of finding the solution. The following is an example of transverse vibrations of an elastic beam to illustrate this point of view.

Let x be the distance measured along the beam. Let u(x,t) be the vertical displacement. Let M(x,t) be the bending moment (about an axis at x perpendicular to the x-axis in the horizontal plane) of the forces acting to the right of x.

(A) We shall first establish by Hooke's law, the fundamental relation for the bending moment.

$$M = \frac{EI}{R} \tag{41}$$

where:

E = Young's modulus

I = The moment of inertia of the cross section at x with respect to the neutral fiber.

R = The radius of curvature at x.

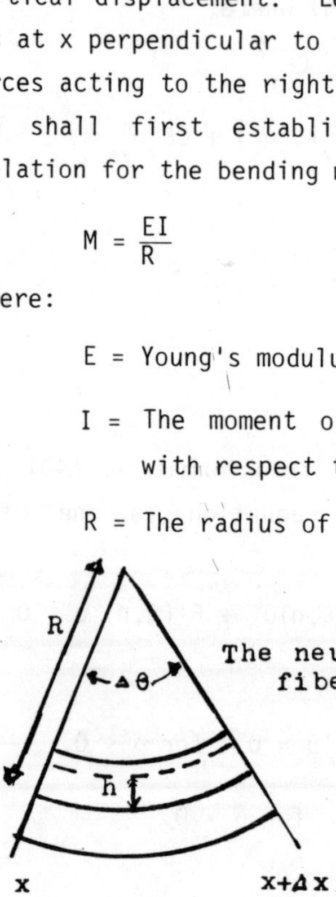

Fig. 3

The cross section at x after the beam is bent, is still assumed to lie in a plane. When the beam is bent upward, the fiber above the neutral one is contracted as indicated in Fig. 3. The force acting on a fiber at a distance h above the neutral fiber is:

$$= E\,\frac{R\Delta\theta - (R-h)\Delta\theta}{R\Delta\theta} = E\,\frac{h}{R}$$

Thus:

$$M = \int_{h_1}^{h_2} Fh\, dh = \frac{E}{R} \int_{h_1}^{h_2} h^2\, dh = \frac{EI}{R} .$$

(B) To find the shear S, let us consider the resulting moment on the element:

$$-M(x,t) + M(x+\Delta x,t) = S\Delta t$$

where S is considered as positive if it is upward. In the limit as $\Delta x \to 0$, we have:

$$\frac{\partial M}{\partial x} = S \tag{42}$$

(C) The moment $M(x,t)$ is balanced with that due to total load $L(x,t)$ to the right of x, which is the sum of the external vertical load $f(\xi,t)$, $(\xi>x)$, and the dynamical load due to the motion of the elements according to Newton's second law. Thus:

$$M(x,t) = \int_x^l (\xi-x)\ L(\xi,t)d\xi + Ax + B$$

where the length of the beam whose initial plane is taken at x = 0. A and B are constants, determined by the concentrated load at the end $x = \ell$. Thus:

$$\frac{\partial^2}{\partial x^2} M(x,t) = L(x,t) = -\rho \frac{\partial^2 u}{\partial t^2} - f(x,t)$$

where ρ is the mass of the beam per unit length, and $f(x,t)$, considered as positive if it is downward, is the load per unit length.

$$\frac{\partial^2}{\partial x^2} \left[\frac{\frac{\partial^2 u}{\partial x^2}}{[1 + (\frac{\partial u}{\partial x})^2]^{3/2}} \right] + \frac{\partial^2 u}{\partial t^2} = -f(x,t)$$

If the deflection is small, the square of $\frac{\partial u}{\partial x}$ can be neglected; and we have:

$$\frac{\partial^2}{\partial x^2} [(EI) \frac{\partial^2 u}{\partial x^2}] + \rho \frac{\partial^2 u}{\partial t^2} = -f(x,t) \tag{43}$$

For a uniform beam, EI is a constant. Therefore, (43) becomes:

$$u_{xxxx} + a^4 u_{tt} = \frac{-f(x,t)}{EI} \tag{44}$$

where:

$$a^4 = \frac{\rho}{EI}$$

7.1 The Boundary Conditions of an Oscillating Beam

The two types of boundary conditions are:

(A) the initial deflection and the initial velocity

(B) the end conditions.

The following three types of end conditions are usually considered:

(B) <u>The end is free</u> -- both the moment of the force and the shear at the end are zero.

$$\frac{\partial^2 u}{\partial x^2} = \frac{\partial^3 u}{\partial x^3} = 0 \tag{45a}$$

(B_2) <u>The end is simply supported or pivoted</u> -- there is no deflection and no moment acting on the end.

$$u = \frac{\partial^2 u}{\partial x^2} = 0 \tag{45b}$$

(B_3) <u>The end is clamped</u> -- there is no deflection and the slope is equal to zero.

$$u = \frac{\partial u}{\partial x} = 0 \tag{45c}$$

7.2 Example

A beam, of unit length, whose ends are simply supported is deflected due to a concentrated load, EIP at its midpoint. If the beam is released from rest and has no more load, the motion of the beam can be found as follows (provided its weight is neglected):

$$u_{xxxx} + a^4 u_{tt} = 0 \tag{46}$$

The boundary conditions are:

(A) $$u(x,0) = f(x) \tag{47}$$

$$u_t(x,0) = 0 \tag{48}$$

(B) $$u(0,t) = u_{xx}(0,t) = 0 \tag{49}$$

$$u(1,t) = u_{xx}(1,t) = 0 \tag{50}$$

The initial deflection can be determined by the theory of a cantilever beam, as in Section 7,+ by solving the following equation with the same end conditions:

$$\frac{d^4u}{dx^4} = -Pe'\left(x-\tfrac{1}{2}\right) \tag{51}$$

where $e(x)$ is Heaviside's unit step function

$$e(x) = \begin{cases} 0, & x < 0 \\ 1, & x > 0 \end{cases}$$

The derivative $e'(x) = \delta(x)$ is called the δ-function, which is a limiting function such that:

$$\int_{-a}^{a} \delta(x)F(x)dx = F(0) .$$

for any positive constant $a > 0$ and any function $F(x)$ continuous at $x = 0$, $\delta(x)$ is therefore not a function in the classical sense and is introduced here merely as an operational device. The limiting case of a concentrated load is kept constant when the area on which the load applies becomes indefinitely small, it leads to the notion of a δ-function. The concept is particularly useful in many fields of

---

+The concept is similar to that of impulse in elementary dynamics.

modern science, however,* a rigorous justification was given only quite recently.** We shall discuss this notion later in our study of potential theory.

The end conditions for (51) are:

$$u(1) = u(0) = u''(0) = u''(1) = 0 .$$

By a simple integration, the solution is found to be:

$$u(x) = f(x) = P[\frac{-1}{16} x + \frac{1}{12} x^3 - \frac{(x-\frac{1}{2})^3}{6} e(x-\tfrac{1}{2})] \qquad (52)$$

Now the elementary solutions of (46) satisfying the homogeneous conditions (48)-(50), by the method of separation of variables, are seen to be:

$$u_n(x,t) = A_n \cos \frac{n^2 \pi x}{a^2} t \sin n\pi x$$

$$(n = 1, 2, 3, ---)$$

Put:

$$u(x,t) = \sum_{n=1}^{\infty} u_n(x,t)$$

Then (47) and (52) give immediately:

$$A_n = 2 \int_0^1 f(x) \sin n\pi x \, dx$$

EXERCISE

By utilizing the facts that:

(i) $$\int x^m e^{ax} dx = e^{ax} [\frac{x^m}{a} - \frac{mx^{m-1}}{a^2} + \frac{m(-1)x^{m-2}}{a^3} - \ldots]$$

(ii) $$\text{Im} \int x^m e^{aix} dx = \int x^m \sin ax \, dx$$

where "m" is a positive integer, "a" is a real constant and "IM" denotes the operation of taking the imaginary part, find

*Especially, the theory of elastic circuits and quantum mechanics.

**L. Schwartz, Theorie des Distributions,Tome 1, 1950 (Hermann and Cle Editeirs).

A_m in Section 7.2.

8. NON-HOMOGENEOUS LINEAR EQUATIONS

The boundary problem of the general non-homogeneous equation in the last section introduces three types of problems: the two types of the problems as indicated in Section 2, and also the following type:

$$L(u) \equiv u_{tt} - M(u) = f(x,t)$$

where M is a linear differential operator of nth order with respect to x only. All the boundary conditions are homogeneous.

$$u(x,0) = u_t(x,0) = 0$$

and:

$$L_i(u) = 0 \quad (i = 1, 2, \text{---}\ n) .$$

The solution of this problem is obtained immediately by quadrature if the problem with the given initial time derivative of u of the homogeneous equation can be solved. Let the solution of the following problem be $u = \varphi(x,t,\tau)$:

$$L(u) = 0$$

$$u(x,\tau) = 0$$

$$u_t(x,\tau) = f(x,\tau)$$

$$L_i(u) = 0 \quad (i = 1, 2, \text{---}\ n)$$

<u>Theorem: if $\varphi(x,t,\tau)$ is the solution of the problem with the prescribed initial time derivative of the homogeneous equation, then:</u>

$$u = \int_0^t \phi(x,t,\tau)\, d\tau$$

<u>is the solution of the non homogeneous equation with all homogeneous boundary conditions.</u>

The operators L_i involve differentiation with respect to the space coordinates only. Thus:

$$L_i(u) = \int_0^t L_i(\varphi)\, d\tau = 0 \qquad (i = 1, \text{----}\ n)$$

that:

$u(x,0) = 0$ is evident.

From:

$$u_t(x,t) = \varphi(x,t,t) + \int_0^t \varphi_t(x,t,\tau)\, d\tau$$

we have:

$$u_t(x,0) = \varphi(x,t,t) = 0$$

Finally,

$$\begin{aligned} L(u) &= u_{tt} - M(u) \\ &= \varphi_t(x,t,t) + \int_0^t L(\varphi) d\tau \\ &= f(x,t) \end{aligned}$$

8.1 The Forced Vibration of a Beam with the Ends Simply Supported

Let a = 1.

$$u_{xxxx} + u_t = f(x,t)$$

The boundary conditions are $u(x,0) = u_t(x,0) = 0$ and:

$$u(0,t) = u_{xx}(0,t) = 0$$

$$u(1,t) = u_{xx}(1,t) = 0$$

By the similar argument as in Section 7.2, we obtain the function:

$$\varphi(x,t,\tau) = \sum_{n=1}^{\infty} A_n(\tau) \sin n^2\pi^2\ (t-\tau) \sin n\pi x\ ,$$

where

$$A_n(\tau) = \frac{2}{h^2\pi^2} \int_0^1 f(x,\tau) \sin n\pi x\, dx$$

then:

$$u(x,t) = \sum_{n=1}^{\infty} T_n(t) \sin n\pi x$$

where:

$$T_n(t) = \frac{2}{n^2\pi^2} \int_0^t d\tau \int_0^1 f(x,\tau) \sin n\pi x \sin n^2\pi^2(t-\tau)\, dx$$

EXERCISE. The Moving Load on a Beam

Find the solution of the problem in Section 8.1, where:

$$f(x,t) = \begin{cases} -p/\alpha & \text{for } ct - \frac{\alpha}{2} \leq x \leq ct + \frac{\alpha}{2} \\ 0 & \text{otherwise.} \end{cases}$$

Find the limiting case as $\alpha \to 0$ and compare with the solution obtained by the operational method using δ-function.

ANSWER.

$$u(x,t) = \frac{4Pc}{x\pi^4} \sum_{n=1}^{\infty} \frac{1}{n^4} \frac{\sin \frac{1}{2} n\pi\alpha}{n^2\pi^2 - c^2} \sin n^2\pi^2 t \sin n\pi x$$

$$- \frac{4P}{\alpha\pi^3} \sum_{n=1}^{\infty} [\frac{1}{n^3} \sin (\frac{n\pi\alpha}{2}) \frac{1}{n^2\pi^2 - c^2} \sin n\pi ct \sin n\pi x.$$

CHAPTER II.

THE METHOD OF ORTHOGONAL FUNCTIONS

For a good many cases of the boundary value problem, it is not only convenient but also imperative to use an appropriate system of coordinates (cf. Ex. 1, Section 1.1). The elementary solutions, obtained by separation of variables, which satisfy all the homogeneous boundary conditions, form a set of orthogonal functions. A linear combination of these functions can, in most cases, represent an arbitrary function in the same sense as the Fourier series.

We shall first deal with the important Sturm-Liouville system of linear differential equations of second order, with two homogeneous boundary conditions and a parameter. The solutions of such a system include many important special functions such as the trigonometric functions, Bessel functions, Legendre functions, etc. Important properties of these functions, as far as they are concerned with the application to the boundary value problem of partial differential equations, are derived in this chapter. Illustrations are given by solving some actual boundary value problems.

1. THE LAPLACIAN OPERATOR IN MUTUALLY ORTHOGONAL CURVILINEAR COORDINATES

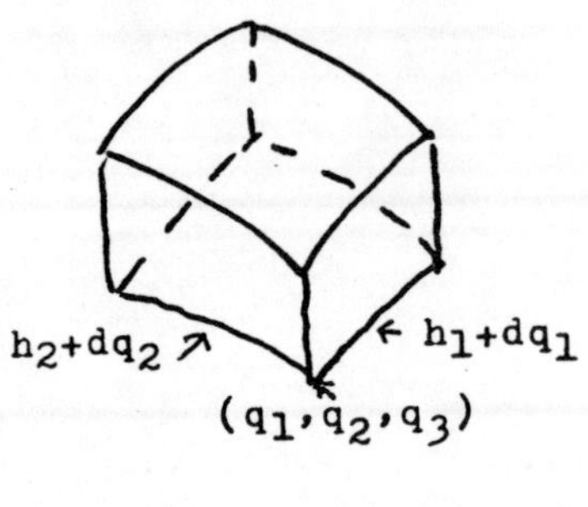

Fig. 4

Let (q_1, q_2, q_3) be the curvilinear coordinates of a point in the space. One can, of course, transform the Laplacian Operator in curvilinear coordinates from the rectangular coordinates. This procedure is rigorous yet it takes some labor in carrying out the calculations. One intuitive way to find the result is by a consideration of an actual physical problem such as the conduction of heat. The Laplacian of the temperature U is the limiting value of the net flux $\bar{F}$ of the heat flowing into an element of volume at a point, per unit volume, provided the conductivity k is unity. Let F_{q_i} be the component of $\bar{F}$ in q_i-direction.

Now let $ds^2 = h_1{}^2 dq_1{}^2 + h_2{}^2 dq_2{}^2 + h_3{}^2 dq_3{}^2$ -- (2). Where s is the element of the arc and h's are functions of q's such that the arc length in q_1 - direction is $h_1 dq_1$, etc. The relation is possible only if the curvilinear system is orthogonal.

Consider the element of the volume bounded by q_i and $q_i + dq_i$ (i = 1, 2, 3). We have:

$$F_{q_1} = \frac{1}{h_1h_2h_3dq_1dq_2dq_3}\left[h_2h_3dq_2dq_3 \left.\frac{\partial U}{h_1\partial q_1}\right|_{q_1+dq_1} - h_2h_3dq_2dq_3\left.\frac{\partial U}{h_1\partial q_1}\right|_{q_1}\right]$$

$$= \frac{1}{h_1h_2h_3}\cdot\frac{\partial}{\partial q_1}\left[\frac{h_2h_3}{h_1}\ \frac{\partial U}{\partial q_1}\right] + \xi_1$$

where: $\varepsilon_1 \to 0$ as $dq_1 \to 0$ (i = 1,2,3)

Similarly we obtain the other two components by the permutation of the subscripts. Taking the limit, we have the Laplacian in the orthogonal curvilinear coordinates:

$$\Delta = \frac{1}{h_1h_2h_3}\left[\frac{\partial}{\partial q_1}\frac{h_2h_3}{\partial h_1}\frac{\partial}{\partial q_1} + \frac{\partial}{\partial q_2}\frac{h_3h_1}{\partial h_2}\frac{\partial}{\partial q_2} + \frac{\partial}{\partial q_3}\frac{h_1h_2}{\partial h_3}\ \frac{\partial}{\partial q_3}\right] \quad (1)$$

1.1 Example. The Cylindrical Coordinates

Let $q_1 = r$, $q_2 = \theta$ and $q_3 = z$

We have:

$$ds^2 = dr^2 + r^2 d\theta^2 + dz^2 \quad (2)$$

and

$$h_1 = 1, \quad h_2 = r, \quad h_3 = 1 \;.$$

Thus, by (1) the Laplacian in the cylindrical coordinates is:

$$\Delta = \frac{1}{r} \frac{\partial}{\partial r} (r \frac{\partial}{\partial r}) + \frac{1}{r^2} \frac{\partial^2}{\partial \theta^2} + \frac{\partial^2}{\partial z^2}$$

EXERCISES

(1) Find the Laplacian in spherical coordinates: $q_1 = r$, $q_2 = \theta$, $q_3 = \varphi$, where r, θ, φ are related to the rectangular coordinates as follows:

$$x = r \sin \varphi \cos \theta$$

$$y = r \sin \varphi \sin \theta$$

$$z = r \cos \varphi$$

ANSWER:

$$\Delta = \frac{1}{r^2} \frac{\partial}{r} (r^2 \frac{\partial}{\partial r}) + \frac{1}{r^2 \sin^2 \varphi^2} \frac{\partial}{\theta^2} + \frac{1}{r^2 \sin\varphi} \frac{\partial}{\partial \varphi} (\sin \varphi \frac{\partial}{\partial \varphi})$$

(2) Ellipsoidal Coordinates

Let q_1, q_2, q_3 be defined by the roots of the following equation:

$$\frac{x^2}{a^2-q} + \frac{y^2}{b^2-q} + \frac{z^2}{c^2-q} = 1 \tag{3}$$

for each set of values of x, y and z, where a, b, c are constants. Then the set (q_1, q_2, q_3) is defined as the ellipsoidal coordinates. Show tht the coordinate axes of the system are mutually orthogonal. Find the Laplacian.

HINT: Show, from the following identity (in q)

$$\frac{x^2}{a^2-q} + \frac{y^2}{b^2-q} + \frac{2}{c^2-q} - 1 \equiv \frac{(q-q_1)(q-q_2)(q-q_3)}{(a^2-q)(b^2-q)(c^2-q)} ,$$

that

$$x^2 = \frac{(a^2-q_1)(a^2-q_2)(a^2-q_3)}{(a^2-b^2)(a^2-c^2)}\ , \quad \text{etc.}$$

From the definition $dx = x_{q_1}dq_1 + x_{q_2}dq_2 + x_{q_3}dq_3$ and etc.

where $x_{q_j} = \frac{\partial x}{\partial q_i}$, derive the general relation:

$$h_i^2 = x_{q_i}^2 + y_{q_i}^2 + z_{q_i}^2 \quad (i = 1,2,3)$$

Then show that:

$$h_1^{-2} = \frac{(a^2-q_1)(b^2-q_1)(c^2-q_1)}{(q_1-q_2)(q_1-q_3)} \quad \text{and so on.}$$

ANSWER:

$$\Delta = \frac{4}{(q_1-q_2)(q_2-q_3)(q_3-q_1)}\ [(q_2-q_3)\{(a^2-q_1)^{\frac{1}{2}}(b^2-q_1)^{\frac{1}{2}}(c^2-q_1)\}^{\frac{1}{2}}\frac{\partial}{\partial q_1}\}^2$$

$$+ (q_3-q_1)\ \{(a^2-q_2)^{\frac{1}{2}}\ (b^2-q_2)^{\frac{1}{2}}\ (c^2-q_2)^{\frac{1}{2}}\ \frac{\partial}{\partial q_2}\}^2$$

$$+ (q_1-q_2)\ \{(a^2-q_3)^{\frac{1}{2}}\ (b^2-q_3)^{\frac{1}{2}}\ (c^2-q_3)^{\frac{1}{2}}\ \frac{\partial}{\partial q_3}\ \}^2]$$

2. A TYPICAL BOUNDARY VALUE PROBLEM. THE CONDUCTION OF HEAT IN A CYLINDER WITH CIRCULAR CROSS SECTION

If the lateral surface of an infinite cylinder is kept at a constant temperature (say zero), the temperature distribution inside has circular symmetry provided the initial distribution is likewise. We have:

$$u_{rr} + \frac{1}{r}\, u_r = u_t \tag{4}$$

The boundary conditions are:

$$u(r,+0) = f(r) \tag{5}$$

and

$$u(L-0,t) = 0 \tag{6}$$

where L is the radius of the cross section. In order that the solution be unique, one more condition has to be imposed such as the solution is bounded as $r \to 0$. This condition is seen later to be equivalent to a boundary condition.

The elementary solutions which satisfy (4) and (5) are:

$$u_\alpha(r,t) = e^{-\alpha^2 t} R_\alpha(r)$$

where $R_\alpha(r)$ is the solution of the following equation:

$$R''_\alpha(r) + \frac{1}{r} R'_\alpha(r) + \alpha^2 R_\alpha(r) = 0 \tag{7}$$

such that $R_\alpha(0)$ is finite and $R_\alpha(L) = 0$.

That α^2 is positive should follow from (7) and (8). However, it can also be seen intuitively from that u_α is bounded for all values of $t > 0$.

The solutions of (7) and (8) are a set of eigenfunctions corresponding to a discrete set of eigenvalues of α. As a matter of fact, these functions except for a constant are Bessel functions of the first kind and of zero order:

$$R_\alpha(r) = J_o(r) \quad .$$

Bessel functions of the second kind do not occur in this problem because of the condition that $R_\alpha(0)$ is finite.* The eigenvalues are determined by:

$$J_o(\alpha L) = 0 \tag{9}$$

---

*From the theory of ordinary linear differential equations, there are two fundamental solutions, such that any solution can be expressed as a linear combination of these two solutions. In this particular case, only one of the fundamental solutions is finite at the singular point $r = 0$.

It is not evident at all that $J_o(r) = 0$ has a discrete set of solutions. This will follow from the general theory of Sturm-Liouville, which we shall study in the next section. In this particular example, we see that equation (7) is unchanged by changing r to -r. Thus, the solution $J_o(r)$ is an even function of r. We consider then only the positive roots $k_n (n = 1, 2, \ldots)$ of $J_o(r) = 0$. Therefore $\alpha_n = k_n/L$. Now let:

$$u(r,t) = \sum_{n=1}^{\infty} A_n e^{-k_n^2 t/L^2} J_o(k_n r/L) \tag{10}$$

In order to satisfy the initial condition (5),

$$f(r) = \sum_{n=1}^{\infty} A_n \varphi_n(r)$$

where:

$$\varphi_n(r) = J_o(k_n r/L) \tag{11}$$

Hence we have now an analogous problem of the expansion of an arbitrary function in terms of the set of the eigenfunctions $\varphi_n(r)$. The possibility of the validity of such an expansion are discussed in the following sections. The present case is so-called the singular case of the Sturm-Liouville system.

The following important property of orthogonality will also be established:

$$\int_0^L r\varphi_n(r)\varphi_m(r)dr = 0 \quad (m \neq n) \tag{12}$$

(cf. Ex. 4).

On account of this orthogonal relation, we are able to derive formally the formula for the generalized Fourier coefficients:

$$A_m = \frac{\int_0^L rf(r)\varphi_m(r)dr}{\int_0^L r\varphi_m^2(r)dr} \tag{13}$$

EXERCISES

(1) Bessel's functions of the first kind

Show, by the integration in series, that (a) the solution of the following equation:

$$L(y) = x^2 \frac{d^2y}{dx^2} + x \frac{dy}{dx} + (x^2-n^2)\, y = 0 \tag{14}$$

(n = integer $\geqq 0$), such that $y(0)$ = finite, is:

$$J_n(x) = (\frac{x}{2})^n \sum_{j=0}^{\infty} \frac{(-1)^j}{j!(n+j)!} (\frac{x}{2})^{2j} \tag{15}$$

except for a constant factor.

HINT: Since $L(x^m) = (m^2-n^2)x^m + x^{m+2}$, we have only to consider a series of the "gap" 2. Thus put:

$$y = x^{\alpha} \sum_{j=0}^{\infty} a_j x^{2j} \tag{15'}$$

Determine α and a_j successively by comparing the coefficients of the same power in x after the series (15') is substituted in (14).

(b) When $n \neq$ an integer (n may be positive or negative), one solution of the Bessel's equation (14) is:

$$J_n(x) = (\frac{x}{2})^n \sum_{j=0}^{\infty} \frac{(-1)^j}{j!\Gamma(n+j+1)} (\frac{x}{2})^{2j} \tag{16}$$

Therefore the general solution is $Aj_n(x) + BJ_{-n}(x)$.

(2) Recursion Formulas

Define $J_{-n}(x) = (-1)^n J_n(x)$ if n = integer. By means of the reduction formula for the Γ- function, $\Gamma(m) = (m-1)\Gamma(m-1)$, show by termwise differentiation of (16), that:

(a) $$2J_n'(x) = J_{n-1}(x) - J_{n+1}(x) \tag{17}$$

(b) $$\frac{2n}{x} J_n(x) = J_{n-1}(x) + J_{n+1}(x) \tag{18}$$

(c) $\frac{d}{dx}[x^n J_n(x)] = x^n J_{n-1}(x)$ (19)

(d) $\frac{d}{dx}[x^{-n} J_n(x)] = -x^{-n} J_{n+1}(x)$ (20)

(3) Show that if $J_n(\lambda L) = 0$ $(n \geqq 0)$ we have:

$$\int_0^L r J_n^2(\lambda r)dr = \frac{L^2}{2}[J_{n+1}(\lambda L)]^2 \tag{21}$$

Thus, in (31), we can evaluate the denominator as:

$$\int_0^L r\varphi_n^2(r)dr = \frac{L^2}{2}[J_{n+1}(\lambda L)]^2 \tag{13'}$$

HINT: Multiply the Bessel's equation:

$$\frac{d}{dr}[r\frac{d}{dr}J_n(\lambda r)] + (\lambda^2 r - \frac{n^2}{r})J_n(\lambda r) = 0 \tag{22}$$

by

$r\frac{d}{dr}J_n(\lambda r)$ and write as:

$$\frac{d}{dr}[r\frac{d}{dr}J_n(\lambda r)]^2 + (\lambda^2 r^2 - n^2)\frac{d}{dr}[J_n(\lambda r)]^2 = 0$$

Integrate this equation and apply the integration by parts to the second term. Then, by (17) and (18), prove (21).

(4) By (22) show that if $J_n(\lambda_1 L) = J_n(\lambda_2 L) = 0$ and $\lambda_1 \neq \lambda_2$,

$$\int_0^L r J_n(\lambda_1 r)J_n(\lambda_2 r)dr = 0$$

2.1 Heat Conduction in a Hollowed Cylinder

Consider the equation

$$u_{rr} + \frac{1}{r}u_r - u_t = 0$$

with the boundary conditions:

$$u(r,0) = f(r) \text{ for } r_1 \leqq r \leqq r_1$$

$$u(r_1,t) = 0$$

$$u(r_2,t) = 0$$

By the method of separation of variables as applied previously the solution is of the form:

$$u(r,t) = \sum_{n=1}^{\infty} A_n e^{-\alpha_n^2 t} \varphi_n(r) \tag{23}$$

where $\varphi_n(r)$ are the eigenfunctions of the following equation:

$$r\varphi'' + \varphi' + \alpha^2 r\varphi = 0 \tag{24}$$

with conditions:

$$\varphi(r_1) = \varphi(r_2) = 0 \; . \tag{25}$$

The fact that the eigenvalues α^2 of the above problem are all positive and form a discrete set follows from the general theory of the Sturm-Liouville system. We will also show that:

$$\int_{r_1}^{r_2} r\varphi_n(r)\varphi_m(r)dr = 0 \tag{26}$$

for $m \neq n$. From the nonhomogeneous boundary condition, it follows:

$$A_n = \frac{\int_{r_1}^{r_2} rf(r)\varphi_n(r)dr}{\int_{r_1}^{r_2} r[\varphi_n(r)]^2 dr} \tag{27}$$

It is worthy of note that the problem can also be interpreted as a process of drying or the diffusion of the concentration of a vapor in a porous cylinder of wood (with a circular ring as cross section) while the dry air is blown along the cylinder.

EXERCISES

(1) The Vibration of a Circular Membrane

Find the solution of the following equation:

$$u_{tt} = u_{rr} + \frac{1}{r} u_r + \frac{1}{r^2} u_{\theta\theta}$$

with the boundary conditions:

$$u(r,\theta,+0) = f(r,\theta)$$

$$u_t(r,\theta,+0) = 0$$

$$u(1,\theta,t) = 0$$

and $u(+0,\theta,t)$ is finite.

ANSWER.

$$u(r,\theta,t) = \sum_{n=1}^{\infty} \sum_{j=0}^{\infty} J_n(\alpha_{n,j} r)(A_{n,j} \cos n\theta + B_{n,j} \sin n\theta) \cos \alpha_{n,j} t$$

$$+ \sum_{j=0}^{\infty} \tfrac{1}{2} A_{o,j}\, J_o(\alpha_{o,j} r) \cos \alpha_{o,j} t$$

where:

$$A_{n,j} = \frac{2}{\pi[J_{n+1}(\alpha_{n,j})]^2} \int_0^1 r J_n(\alpha_{n,j} r) dr \int_{-\pi}^{\pi} f(r,\varphi) \cos n\varphi \, d\varphi$$

$$B_{n,j} = \frac{2}{\pi[J_{n+1}(\alpha_{n,j})]^2} \int_0^1 r J_n(\alpha_{n,j} r) dr \int_{-\pi}^{\pi} f(r,\varphi) \sin n\varphi \, d\varphi$$

and:

$$J_n(\alpha_{n,j}) = 0 (\alpha_{n,j} > 0) \quad \text{for} \quad n,j = 0, 1, \ldots$$

(2) Heat Conduction in a Thin Circular Plate

Find the temperature $u(r,t)$ in a thin plate whose edge $r = 1$ is kept at the temperature $u = 0$, and whose initial temperature on the plate is $u = 1$, when there is surface heat transfer (by radiation) from the circular faces to surrounding medium at zero temperature. The equation for the heat condiction is:

$$u_t = u_{rr} + \frac{1}{r} u_r - hu$$

where the conductivity is assumed to be unity and h is a positive constant.

ANSWER.

$$u(r,t) = 2e^{-ht} \sum_{n=1}^{\infty} e^{-\alpha_n^2 t} J_0(\alpha_n r)/\alpha_n J_1(\alpha_n)$$

where:

$$\alpha_n > 0 \quad \text{and} \quad J_0(\alpha_n) = 0 \; .$$

(3) Bessel's Functions of the Second Kind of Order Zero

(a) Show that a second solution of the equation:

$$L(y) \equiv xy'' + y' + xy = 0$$

is

$$\frac{\pi}{2} Y_0 = [\log \gamma(\tfrac{x}{2})]J_0(x) - \sum_{j=1}^{\infty} \frac{(-1)^j}{(j!)^2} (1+\tfrac{1}{2}+\ldots+\tfrac{1}{j}) (\tfrac{x}{2})^{2j}$$

where

$$\log \gamma = \lim_{m\to\infty} (\sum_{j=1}^{m} \frac{1}{j} - \log m) = 0.577215 \ldots$$

HINT: Show that $L[y(\alpha,x)] \equiv a_0\alpha^2 x^{\alpha-1}$

where:

$$y(\alpha,x) = x^{\alpha} \sum_{j=0}^{\infty} a_j(\alpha)x^{2j} \quad \text{provided}$$

$$a_{j+1} = \frac{1}{(2n+2j+\alpha)^2} a_j = (-1)^{j+1} \frac{1}{(2n+2j+\alpha)^2 \ldots (2+\alpha)^2} a_j$$

Then by the identity:

$$\frac{\partial}{\partial\alpha} L[y(\alpha,x)] \equiv L[\frac{\partial y}{\partial\alpha}] \equiv \alpha(2x^{\alpha-1}+\alpha x^{\alpha-1} \log x) a_0$$

it follows that: $\left.\frac{\partial y}{\partial \alpha}\right|_{\alpha=0}$ is a solution of the Bessel's equation of order zero.

(b) Show that the eigenfunctions defined by (24) and (25) are, except a constant factor,

$$\varphi_n(r) = Y_0(\alpha r_1)J_0(\alpha r) - J_0(\alpha r_1)Y_0(\alpha r)$$

where α is a positive root of the following equation:

$$\begin{vmatrix} J_0(\alpha r_1), & Y_0(\alpha r_1) \\ J_0(\alpha r_2), & Y_0(\alpha r_2) \end{vmatrix} = 0$$

3. STURM-LIOUVILLE SYSTEMS

The eigenvalue problems of the last section are special cases of the following system:

$$L(y) + \lambda r(x)y \equiv \frac{d}{dx}[p(x)\frac{dy}{dx}] + [q(x) + \lambda r(x)]\cdot y = 0^* \qquad (27)$$

with the added conditions:

$$L_1(y) \equiv Ay(a) + By'(a) = 0$$
$$L_2(y) \equiv Cy(b) + Dy'(b) = 0 \qquad (28)$$

where A, B, C and D are prescribed real constants such that $(A,B) \neq (0,0)$ and $(C,D) \neq (0,0)$. We always impose the assumptions that:

(A) $p(x)$, $p'(x)$, $q(x)$ and $r(x)$ are continuous in $[a,b]$.

(B) $p(x)$ has the same sign throughout the interval $[a,b]$.

Without loss of generality it can be assumed that $p(x) \geq 0$ in $[a,b]$ $\cdot p(x) = 0$ only if $x = a$ or $x = b$. The case that $p(a) = 0$ or $p(b) = 0$ is called the singular case. In the latter case, only one equation of (28) is required. A separate treatment of the singular

---

*Any linear differential equation of the second order can be reduced to this form by a simple transformation of the dependent variable.

case is required. Equation (27), with the conditions (29), has the trivial solution $y = 0$.

Definition 1. A solution $y = y(x)$ other than the trivial one is called an eigenfunction. As we have seen from the preceding examples, the eigenfunctions exist only for some particular values of λ.

Definition 2. A value of λ, for which a non-trivial solution exists, is called an eigenvalue.

Definition 3. Two functions $\varphi_m(x)$ and $\varphi_n(x)$ are said to be orthogonal with respect to the weight function $r(x)$ if:

$$\int_a^b r(x)\varphi_m(x)\varphi_n(x)dx = 0 \tag{29}$$

3.1 Orthogonality of Eigenfunctions:

Theorem 1. If λ_m and λ_n are two distinct eigenvalues and if $\varphi_m(x)$ and $\varphi_n(x)$ are the corresponding eigenfunctions, then they are orthogonal with respect to the weight function $r(x)$.

Proof. From (28), we have:

$$A\varphi_m(a) + B\varphi_m{}'(a) = 0$$

and

$$A\varphi_n(a) + B\varphi_n{}'(a) = 0$$

Since $(A,B) \neq (0,0)$, it follows that:

$$\begin{vmatrix} \varphi_m(a), & \varphi_m{}'(a) \\ \varphi_n(a), & \varphi_n{}'(a) \end{vmatrix} = 0 \tag{30}$$

Similarly, we derive the same relation at $x = b$. Now,

$$L[\varphi_m(x)] + \lambda_m r(x)\varphi_m(x) = 0$$

and

$$L[\varphi_n(x)] + \lambda_n r(x)\varphi_n(x) = 0 \qquad (31)$$

Then it is clear that:

$$(\lambda_m - \lambda_n) r(x)\varphi_m(x)\varphi_n(x)$$

$$= \varphi_m(x)L[\varphi_n(x)] - \varphi_n(x)L[\varphi_m(x)]$$

$$= \frac{d}{dx}[\varphi_m(x)\cdot p(x)\varphi_n'(x) - \varphi_n(x)\cdot p(x)\varphi_m'(x)] \qquad (32)$$

Integrate (32) from $x = a$ to $x = b$ and apply (30). The relation (29) follows directly.

3.2 Reality of Eigenvalues

<u>Theorem 2.</u> <u>If $r(x)$ is not identically equal to zero and has the same sign throughout the interval $[a,b]$ then all the eigenvalues of the Sturm-Liouville system are real.</u>

<u>Proof.</u> Let $\lambda = \alpha + i$ be the eigenvalue and let $\varphi(x) = u(x) + iv(x)$ be the corresponding eigenfunction. Substitute in (27) and separate the real and the imaginary parts, and it follows:

$$L(u) + r(x)[\alpha u(x) - \beta v(x)] = 0 \qquad (33)$$

and:

$$L(v) + r(x)[\alpha v(x) + \beta u(x)] = 0 \qquad (34)$$

Consequently, we have:

$$vL(u) - uL(v) - \beta r(x)\,[u^2(x) + v^2(x)] = 0 \qquad (35)$$

Integrating (35), it is evident by means of (32) that:

$$\beta \int_a^b r(x)(u^2 + v^2)dx = 0$$

since both u and v satisfy (28) and (30) separately. Thus $\beta = 0$,

3.2.1. Some special cases. The constancy of the sign of eigenvalues.

If in addition to the assumption that r(x) has the same sign throughout (a,b), $q(x) \leq 0$ in (a,b) then it is possible to extend the result of the last section and conclude that all the eigenvalues are not only real, but also have the same sign as r(x) under the following boundary conditions:

(A) $\varphi(a) = 0$ or $\varphi'(a) = 0$

(B) $\varphi(b) = 0$ or $\varphi'(b) = 0$

Proof.

$$\lambda \int_a^b r(x)\varphi^2(x)dx = \int_a^b [-\varphi \frac{d}{dx}(p\varphi')-q\varphi^2]dx$$

$$= -p(x)\varphi(x)\varphi'(x) \Big|_a^b + \int_a^b (p\varphi'^2-q\varphi^2)dx$$

from which the result follows directly.

EXERCISES

(a) Show that all the eigenvalues in the examples of Section 2 and Section 2.1 are positive.

(b) Show that the problem:

$$\frac{d^4y}{dx^4} + [q(x)+\lambda r(x)]\, y = 0$$

with the conditions $y(0) = y(1) = 0$, $y'(0) = y'(1) = 0$ has its eigenvalues of opposite sign from r(x) if $q(x) \geqq 0$.

3.3 Simplicity of the Eigenvalue

Theorem 3. Corresponding to each eigenvalue there is one and only one linearly independent eigenfunction of a non-singular Sturm-Liouville system.

Proof. It is sufficient to show that if $\varphi(x)$ is an eigenfunction corresponding to λ, then every other eigenfunction corresponding to the same eigenvalue λ is $C\varphi(x)$ where C is a constant. Suppose that this is false. Let $\psi(x)$ be an other eigenfunction linearly independent of $\varphi(x)$ corresponding to the same λ. Let $p(a) \neq 0$. By the elementary theory of linear differential equation any solution of the differential equation is a linear combination of $\psi(x)$ and $\varphi(x)$. This is impossible since a general solution will not satisfy the condition (28).

One can apply this result to the example in Section 2.1 and conclude that $J_0(x)$ and $Y_0(x)$ cannot vanish simultaneously.

4. GENERALIZED FOURIER EXPANSIONS

The problem of the existence of eigenvalues and the expansion of an arbitrary function in terms of eigenfunctions cannot be treated without further knowledge of analysis. The following expansion is found very useful in solving the boundary value problem. We shall establish it rigorously in the Appendix.

Theorem. If $(pr)''$ is continuous* and $r(x)$ is positive and if assumptions (A) and (B) of Section 2 hold, then the eigenvalues of the non-singular Sturm-Liouville system are all real, and there are at most a finite number of them which are negative. Let λ_n and $\varphi_n(x)$ be corresponding eigenvalues and eigenfunctions. If f(x) is continuous and of bounded variation** in [a,b] then:

---

*This condition is required in the proof in the Appendix. It is not necessary, especially for the conclusion in the first part of the theorem.

**A function f(x) is said to be "of bounded variation" in [a,b] if:

$$\sum_{i=1}^{m} |f(x_i)-f(x_{i-1})| < A \text{ where } [x_0 = \alpha < x_1 < x_2 \dots < x_{m-1} < x_m = b]$$

is a subdivision of [a,b] and A is a constant, independent of the number and the way of the subdivision. A function satisfying the Dirichlet conditions (A) and (B) in Section 3 of Chapter I is evidently of bounded variation.

$$f(x) = \sum_{n=1}^{\infty} A_m \varphi_n(x) \tag{36}$$

where:

$$A_n = \int_a^b f(x) r(x) \varphi_n(x) dx / \int_a^b r(x) \varphi_n^2(x) dx \tag{37}$$

The series (36) converges uniformly in any closed subinterval of the interval of [a,b]. For the singular case, a similar expansion holds. It has to be treated separately.

4.1 Example. The Steady Flow of Temperature in an Infinite Prism

Consider the boundary problem where:

$$u_{xx} + u_{yy} = 0 \tag{38}$$

with the conditions:

$$u(0,y) = 0 \tag{39}$$

$$u_x(1,u) + hu(1,y) = 0 \tag{40}$$

$$u(x,0) = 0 \tag{41}$$

$$u(x,1) = 1 \tag{42}$$

Let us first look for the elementary solutions of the form X(x)• Y(y) such that (39), (40), and (41) are satisfied. We have:

$$X'' + \lambda X = 0; \quad X(0) = 0; \quad \text{and} \quad X'(1) + hX(1) = 0 \tag{43}$$

and

$$Y'' + \lambda Y = 0; \quad Y(0) = 0 \tag{44}$$

From the theorem in Section 3.2, λ must be positive. Thus, put $\lambda = \alpha^2$. It follows then from (43) and (44) that:

$$Y = A \sin h(\alpha_n y)$$

$$X = B \sin(\alpha_n x)$$

where α_n are the positive roots of the equation:

$$\alpha \cos \alpha + h \sin \alpha = 0 \quad \text{or} \quad \frac{1}{h}\alpha = -(\tan \alpha). \tag{45}$$

The existence of a discrete set of roots of (45) is seen from the following Fig. 5.

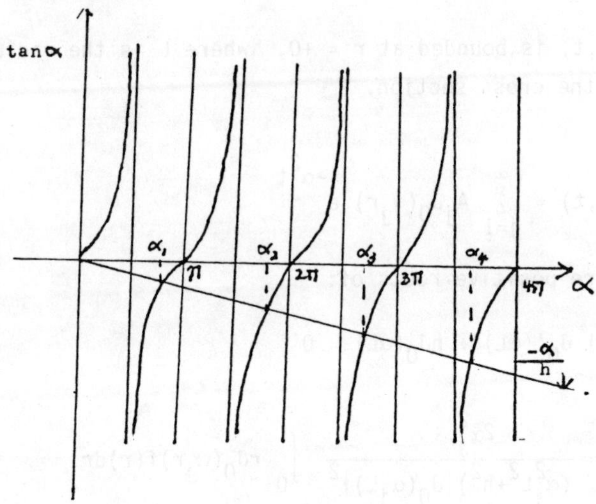

Fig. 5

Therefore, we have:

$$u(x,y) = \sum_{n=1}^{\infty} A_n \sin h\,(\alpha_n y) \sin (\alpha_n x)$$

where A_n, which is obtained by means of (42) and (37), is found to be:

$$A_n = \frac{2h}{\alpha_n \sin h(\alpha_n)} \frac{1 - \cos \alpha_n}{h + \cos^2 \alpha_n} \tag{46}$$

EXERCISE

Heat Conduction in an Infinite Circular Cylinder with Radiation at the Surface

Solve the following equation:

$$u_t = u_{rr} + \frac{1}{r} u_r$$

with the conditions:

$$u(r,+0) = f(r)$$

$$Lu_r(L,t) + hu(L,t) = 0$$

and

u(r,t) is bounded at r = +0, where L is the radius of the cross section.

ANSWER.

$$u(r,t) = \sum_{j=1}^{\infty} A_j J_0(\alpha_j r)\, e^{-\alpha_j^2 t}$$

where α_j are positive roots of:

$$(\alpha L)\, J_0{}'(\alpha L) + hJ_0(\alpha L) = 0$$

and

$$A_j = \frac{2\alpha_j^2}{(\alpha_j^2 L^2 + h^2)\{J_0(\alpha_j L)\}^2} \int_0^L rJ_0(\alpha_j r) f(r) dr$$

(j = 1, 2, ...). Show that if the surface is insulated, that is h = 0, $\alpha_0 = 0$ and:

$$A_0 = \frac{2}{L^2} \int_0^L rf(r)\, dr\ .$$

5. SINGULAR STURM-LIOUVILLE SYSTEMS

The theorems concerning orthogonality and reality of eigenfunctions and eigenvalues in Section 3 are valid for the same singular cases when p(a) = 0 or p(b) = 0. They are valid even when p(a) = p(b)= 0, provided that instead of the boundary conditions, the eigenfunctions together with their first derivatives are required to be continuous in [a,b] (cf. (32) and (35)). The eigenfunctions of these systems occur very often in solving boundary problems of partial differential equations. As an illustration let us consider the problem of the potential V in the space due to a fixed distribution of electric potential $V = f(\varphi)$ on a unit sphere where r, φ, θ are the ordinary spherical coordinates. The space is assumed to be free of electric charges.

That the electric potential satisfies the Laplace equation $\Delta V=0$ is derived in the same way as the equation of heat conduction since the electric field in a direction is defined as the negative directional derivative of the potential. In the present case, we are looking for the solution which is independent of θ. Thus, we have the equation in spherical coordinates:

$$\frac{1}{r}\frac{\partial^2}{\partial r^2}(rV) + \frac{1}{\sin\varphi}\frac{\partial}{\partial\varphi}(\sin\varphi\frac{\partial V}{\partial\varphi}) = 0 \tag{47}$$

The boundary conditions:

$$\lim_{r\to 1} V(r,\varphi) = f(\varphi) \tag{48}$$

and

$$\lim_{r\to\infty} V(r,\varphi) = 0 \tag{49}$$

$V(r,\varphi)$ and its partial derivative with respect to φ are bounded. The method of separation of variables gives the elementary solutions of the form:

$$R(r)\cdot\Phi(\varphi)$$

where:

$$r^2R'' + 2rR' - \lambda R = 0 \tag{50}$$

and

$$\frac{1}{\sin\varphi}\frac{d}{d\varphi}\{\frac{1-\cos^2\varphi}{\sin\varphi}\}\frac{d\Phi}{d\varphi} + \lambda\Phi = 0 \tag{51}$$

or

$$\frac{d}{dx}\{(1-x^2)\frac{d\Phi}{dx}\} + \lambda\Phi = 0 \tag{51'}$$

where $X = \cos\varphi$. We are looking for the solutions of (51'), which together with its first derivatives, are bounded at $x = +1$. <u>We shall see in the section which immediately follows that $\lambda = n(n+1)$ where $n = 1,2,\ldots$.</u>

The corresponding eigenfunctions, except for a constant factor, are Legendre polynomials:

$$P_n(x) = \frac{1}{2^n n!} \frac{d^n}{dx^n} (x^2-1)^n \quad .$$

Now, the general solution of (50) is:

$$R = Ar^n + Br^{-(n+1)} \quad .$$

For $r \leqq 1$, since R is bounded, B = 0. We have:

$$V(r,\varphi) = \sum_{n=0}^{\infty} A_n r^n P_n(\cos\varphi)$$

where:

$$A_n = \frac{\int_{-1}^{1} g(x)P_n(x)dx}{\int_{-1}^{1} P_n^2(x)dx} \tag{52}$$

and:

$$g(x) = f(\cos^{-1} x)$$

by means of (48) and the orthogonality of $P_n(x)$. The expansion is possible if g(x) satisfies certain conditions, such as those stated in the theorem of Section 4.

Similarly, for $r \geqq 1$, the potential:

$$V(r,\varphi) = \sum_{n=0}^{\infty} A_n r^{-(n+1)} P_n(\cos \varphi) \quad .$$

5.1 Legendre Polynomials

Consider the equation:

$$(1-x^2) \frac{d^2y}{dx^2} - 2x \frac{dy}{dx} + n(n+1)y = 0 \quad . \tag{53}$$

For a non-negative integral value of n, the equation has a polynomial solution by the method of integration in series:

$$P_n(x) = \frac{1}{2^n} \sum_{j=0}^{m} \frac{(-1)^j (2n-2j)!}{j!(n-j)!(n-2j)} x^{n-2j} \tag{54}$$

where $m = [\frac{n}{2}]$ = the integral part of $\frac{n}{2}$. It can also be verified directly that:

$$P_n(x) = \frac{1}{2^n n!} \frac{d^n}{dx^n} (x^2-1)^n \tag{55}$$

From the discussions in Section 5, these Legendre polynomials form an orthogonal set. Here we have manifestly shown that $\lambda = n(n+1)$ are the eigenvalues of the singular Sturm-Liouville system (51'). However, we have not shown that they are the only eigenvalues. In order to establish the last assertion, we are going to prove the following:

Theorem: If a function with a continuous derivative in $[-1,1]$ is orthogonal to $P_n(x)$ for $n = 0,1,2, \ldots$, the function is identically zero.

Corollary. For each $\lambda = n(n+1)$, the singular system (51') has only one linearly independent eigenfunction, namely $P_n(x)$.

Suppose $Q_n(x)$ be another eigenfunction corresponding to $\lambda = n(n+1)$ and linearly independent of $P_n(x)$. Set $Q_n(x) = \bar{Q}_n(x) + AP_n(x)$, $\bar{Q}_n \not\equiv 0$, where:

$$A = -\int_{-1}^{1} P_n Q_n dx \Big/ \int_{-1}^{1} P_n^2(x) dx ,$$

is also linearly independent of $P_n(x)$. Then:

$$\int_{-1}^{1} \bar{Q}_n P_m dx = 0 \quad \text{for} \quad m = 0, 1, 2, \ldots$$

Thus $\bar{Q}_n(x) \equiv 0$ by the theorem. This is a contradiction.

Proof of the Theorem.

Since $P_n(x)$ is a polynomial of degree n, it is clear that for any integer $m \geq 0$, x^m can be expressed as a linear combination of

($P_0(x)$, $P_1(x)$, ... $P_m(x)$. Thus if:

$$\int_{-1}^{1} f(x)P_m(x)dx = 0 \quad \text{for} \quad m = 0, 1, \ldots,$$

then:

$$\int_{-1}^{1} f(x)x^m dx = 0 \quad .$$

For a fixed n, both sin nπx and cos nπx can be expanded as uniformly convergent power series. Consequently, we have:

$$\int_{-1}^{1} f(x) \sin n\pi x \, dx = 0$$

$$\int_{-1}^{1} f(x) \cos n\pi x dx = 0 \quad \text{for} \quad n = 0, 1, \ldots .$$

Now $f(x)$ is continuously differentiable, therefore, it is of bounded variation. Because the variation of $f(x)$:

$$\text{Var. } f(x) \text{ in } [1,-1] \leq \int_{-1}^{1} |f'(x)| \, dx$$

and the theorem in Section 4, the Fourier series of $f(x)$ (which is always zero) converges to the function $f(x)$ at every point in the interior of [-1,1]. Thus $f(x) \equiv 0$.

<u>EXERCISES</u>: <u>Elementary Properties of Legendre Polynomials</u>.

(1) <u>Recursion Formula</u>

Show that:

$$(n+1)P_{n+1}(x) - x(2n+1)P_n(x) + nP_{n-1}(x) = 0 \; . \tag{56}$$

(2) <u>Reality of the Roots</u>

Show by Rolle's theorem that all roots of $P_n(x)$ are real and lie between -1 and 1.

(3) <u>Orthogonality</u>

If $m < n$ and $Q(x)$ is a polynomial of degree m, show that:

$$\int_{-1}^{1} P_n(x)Q(x)dx = 0 .$$

This property characterizes $P_n(x)$ up to a constant factor.

(4) Normalization

Set $\varphi_n(x) = \sqrt{n+\frac{1}{2}}\, P_n(x)$

Show that:

$$\int_{-1}^{1} \varphi_n^2(x)dx = 1 .$$

(5) Generating Functions

Show that:

$$(1-2xz+z^2)^{-\frac{1}{2}} = \sum_{m=0}^{\infty} P_n(x)z^n \tag{57}$$

for $|x| \leqq 1$ and $|z| < 1$. Hence, $P_n(1) = 1$.

HINT. Differentiate with respect to z and multiply by $(1-2\,xz+z^2)$. Then compare the coefficients of z^n. Use the recursion formula.

5.1.1 The expansion in Legendre polynomials. The same expansion theorem as in Section 4 is valid if the set of eigenfunctions is the set of Legendre polynomials. As a matter of fact, the theorem can be generalized for other orthogonal polynomials which are to be dealt with later. We shall make use of this fact.

EXERCISE. Gravitational Potential due to a Circular Plate

$$\frac{1}{r}\frac{\partial^2}{\partial r^2}(rV) + \frac{1}{\sin\varphi}\frac{\partial}{\partial\varphi}(\sin\varphi\,\frac{\partial V}{\partial\varphi}) = 0 \tag{58}$$

with the boundary conditions:

$$V(r,0) = k\int_0^1 \frac{2\pi x}{(r^2+x^2)^{\frac{1}{2}}}\,dx = 2\pi k\,(\sqrt{r^2+1} - r) \tag{59}$$

$V(r,\varphi)$ and $V_\varphi(r,\varphi)$ are bounded.

ANSWER.

$$V(r,\varphi) = 2\pi k\ [1-rP_1\ (\cos\varphi) + \tfrac{1}{2}\, r^2P_2(\cos\varphi)$$

$$- \frac{1}{2\cdot4}\, r^4P_4(\cos\varphi) + \frac{1\cdot3}{2\cdot4\cdot6}\, r^6P_6\ (\cos\varphi) + \ldots$$

for $0 < r < 1, \quad 0 \leqq \varphi \leqq \pi/2$.

$$V(r,\varphi) = 2\pi k\ [\frac{1}{2}\frac{1}{r} - \frac{1}{2\cdot4}\cdot\frac{1}{r^3}\, P_3(\cos\theta) + \ldots]$$

for $r > 1$.

6. OTHER ORTHOGONAL POLYNOMIALS

The following sets of orthogonal polynomials are further examples of eigenfunctions of singular Sturm-Liouville systems. They can be treated exactly the same way as for Legendre polynomials.

(i) Tchebycheff Polynomials.

The weight function is:

$$r(x) = \frac{1}{(1-x^2)^{\frac{1}{2}}}\ ; \quad a = -1, \quad b = +1\ .$$

The differential equation is:

$$(1-x^2)y'' - xy' + y = 0 \tag{61}$$

or:

$$\frac{d}{dx}\,[(1-x^2)^{\frac{1}{2}}\, y'] + \frac{\lambda}{(1-x^2)^{\frac{1}{2}}}\, y = 0 \tag{61'}$$

The eigenfunctions, which have continuous derivatives in [-1,1] are:

$$T_n(x) = \frac{1}{2^{n-1}}\cos\,(n\cos^{-1}x)\ (n = 0,1,2,\ \ldots) \tag{62}$$

corresponding to eigenvalues $\lambda = n^2$.

(ii) Jacobi Polynomials.

The weight function is:

$$r(x) = x^p(1-x)^q \quad \text{where} \quad p,q > -1 \quad \text{and} \quad a = 0, \quad b = 1\ .$$

The differential equation is:

$$x(1-x)y'' + [p+1 -x(p+q+2)]y' + \lambda y = 0 \quad (63)$$

or:

$$\frac{d}{dx}[x^{p+1}(1-x)^{q+1}y'] + \lambda x^{p}(1-x)^{q}y = 0 \quad (63')$$

The eigenfunctions are:

$$G_n(q,p+1,x) = \frac{x^{-p}(1-x)^{p+1-q}}{(p+1)\,(p+n)}\frac{d^n}{dx^n}[x^{p+n}(1-x)^{q+n-p-1}] \quad (64)$$

Corresponding the eigenvalues $\lambda = (p+q+1+n)n$, $(n = 0,1,2,\ldots)$.

It is worhty of note that the Legendre polynomials and the Tchebycheff polynomials are special cases of the Jocobi or the hyper-geometric polynomials:

$$G_n(1,1,\frac{1-x}{2}) = P_n(x)$$

$$\frac{1}{2^{n-1}}G_n\,(0,\frac{1}{2},\frac{1-x}{2}) = T_n(x)$$

Further information and elementary properties of these polynomials and other special functions can be found in the following book:

"Formulas and Theorems for the Special Functions of Mathematical Physics," by W. Magnus and F. Oberhettinger (1949).

6.1 Singular Sturm-Liouville Systems for the Infinite Interval

The case of the singular Sturm-Liouville systems when one or both of the end points a and b become infinite are of great importance in classical mathematical physics. The following are two examples:

(i) Laguerre Polynomials

The weight function is $r(x) = e^{-x}$; $a = 0$ and $b = \infty$. The differential equation is:

$$xy'' + (1-x)y' + \lambda y = 0 \quad (65)$$

or

$$\frac{d}{dx}(xe^{-x}y') + \lambda e^{-x}y = 0 \tag{65'}$$

The eigenfunctions are defined as those non-trivial solutions of (65') such that:

(A) y and y' are bounded at x = +0.

(B) $\sqrt{x}\, e^{-x/2}\, y$ and $\sqrt{x}\, e^{-x/2}\, y'$ are continuous and integrable over the interval $(0, \infty)$; i.e.

$$\int_0^\infty e^{-x}\, |y|\, dx < \infty .$$

Thus, the orthogonality of eigenfunctions over the interval $(0, \infty)$, the reality and positivity of the eigenvalues are established as before. Also, the same argument may be applied to show that $\lambda = n$ (n = 0,1,2 ...) are eigenvalues and the eigenfunctions are:

$$e^{-\frac{x}{2}} L_n(x)$$

where:

$$L_n(x) = e^x \frac{d^n}{dx^n}(x^n e^{-x}) \tag{66}$$

are the Laguerre polynomials. To show that these are the only eigenvalues, we use the fact that any continuous function f(x) such that:

$$\int_0^\infty f^2(x)\, e^{-x} dx < \infty$$

and f(x) is orthogonal to all $L_n(x)$ with respect to the weight function e^{-x} must vanish identically in $[0, \infty]$. The proof of this assertion is given in the example of the next section.

(ii) Hermite Polynomials

The weight function is $r(x) = e^{-x^2}$. $a = -\infty$, $b = +\infty$. The differential equation is:

$$y'' - 2xy' + \lambda y = 0 \tag{67}$$

or

$$\frac{d}{dx}(e^{-x^2}y') + \lambda e^{-x^2} y = 0 \tag{67'}$$

The eigenfunctions are the solutions of (67') such that:

(A) $e^{-\frac{x^2}{2}} y$ and $e^{-\frac{x^2}{2}} y' \to 0$ as $x \to \pm\infty$

and

(B) $\int_{-\infty}^{\infty} e^{-x^2} y^2 \, dx < \infty$

It is exactly the same reasoning as above the eigenvalues are shown to be $\lambda = 2n$ ($n = 0,1,2,\ldots$) and the corresponding eigenfunctions, except for a constant factor, are:

$$H_n(x) = (-1)^n e^{x^2} \frac{d^n}{dx^2} (e^{-x^2}) \tag{68}$$

The orthogonality of the eigenfunctions follows from the general theory of Sturm-Liouville system.

6.2 Example. The Harmonic Oscillator

Consider the Schrodinger's equation:

$$u_{xx} + u_{yy} + u_{zz} + \frac{8\pi^2 m}{h^2}(E-V)u = 0 \tag{69}$$

The equation has the physical interpretation that u^2 is the probability per unit volume for an electron to be found at a point in the space and that E = the total energy of an electron, V = the potential of a field, m = the mass of an electron and h = Planck's constant.

Consider now an elastic field:

$$V = 2\pi^2 m\,(\nu_1^2 x^2 + \nu_2^2 y^2 + \nu_3^2 z^2) .$$

We are looking for a solution in the form:

$$u = X(x)\ Y(y)\ Z(z)$$

Then:

$$X'' + \frac{8\pi^2 m}{h^2}(E_1 - 2\pi^2 m\nu_1^2 x^2)X = 0$$

where E_1 is a constant interpreted as the energy which can be released. The conditions for X are:

(A) X and $X' \to 0$ as $x \to \pm\infty$

and

$$\text{(B)} \quad \int_{-\infty}^{\infty} X^2\, dx < \infty\ . \tag{69}$$

These are two similar equations for Y and Z. Put

$$\xi = \left[\frac{4\pi^2 m\nu_1}{h}\right]^{\frac{1}{2}} x$$

and

$$X = e^{-\xi^2/2}\ H(\xi).$$

We have:

$$\frac{d^2H}{d\xi^2} - 2\xi\,\frac{dH}{d\xi} + 2\left[\frac{E_1}{h\nu_1} - \frac{1}{2}\right] H = 0$$

which is the equation (67). Because of (69), the eigenvalues are:

$$2\left[\frac{E_1}{h\nu_1} - \frac{1}{2}\right] = 2n \quad (n = 0,1,\ldots)$$

or:

$$E_1 = (n + \tfrac{1}{2})\ h\nu_1\ ,$$

that is to say, the energy levels are quantized.

EXERCISES

(1) Elementary Properties of Laguerre Polynomials .

Show that:

(i) $L_{n+1}(x) - (2n+1-x)L_n(x) + n^2L_{n-1}(x) = 0$

$xL_n'(x) = nL_n(x) - n^2L_{n-1}(x)$

(ii) $\int_0^\infty e^{-x} L_n^2(x)\, dx = (n!)^2$

(iii) $\frac{1}{1-t} e^{-\frac{tx}{1-t}} = \sum_{n=0}^{\infty} \frac{t^n}{n!} L_n(x), \quad |t| < 1$

<u>HINT</u>: Differentiate with respect to t and compare the coefficients of the same power in t. Then use (i).

(2) <u>Elementary Properties of Hermite Polynomials</u>

(i) $H_{n+1}(x) - 2xH_n(x) + 2nH_{n-1}(x) = 0$

$H_n'(x) = 2nH_{n-1}(x)$

(ii) $\int_{-\infty}^{\infty} e^{-x^2} H_n^2(x)\, dx = 2^n n!(\pi)^{\frac{1}{2}}$

(iii) $e^{-t^2+2tx} = \sum_{n=0}^{\infty} \frac{H_n(x)}{n!} t^n$

7. GENERAL ORTHOGONAL FUNCTIONS

The foregoing examples are special cases of a general theory. One can extend the results of a much broader class of functions, which is a realization of <u>Hilbert Space</u>. The idea of convergence has to be extended to what is called "convergence in the mean." The following few pages are devoted to some important notions which are very often employed in <u>generalized Fourier analysis.</u>

Consider* the class K_2 of all the functions (x) such that:

*The class K_2 considered in this section may either be the class R_2 of all functions square-integrable in Riemann's sense or the class L_2 of all functions square-integrable in Lebesgue's sense. It should be remarked that R_2 does not form a Hilbert space.

$$\int_a^b \varphi^2(x)\, dx \quad < \infty ,$$

where one or both of the limits a and b may be infinity. This hypothesis is imposed throughout this section.

An important property of this class of functions is that any (finite) linear combination of the functions still belongs to this class.

7.1 Schwarz's Lemma

The following lemms plays a fundamental role. <u>If $\varphi(x)$ and $\psi(x)$ are both square integrable, then:</u>

$$\int_a^b |\varphi(x)\psi(x)|\, dx \leqq [\int_a^b \varphi^2\, dx \cdot \int_a^b \psi^2\, dx]^{\frac{1}{2}} \tag{70}$$

<u>Proof.</u> Since, for any A and B

$$\int_a^b \{A|\varphi(x)| + B|\psi(x)|\}^2 \quad dx \geqq 0$$

then:

$$\{\int_a^b \varphi^2 dx\}\ A^2 + 2\{\int_a^b |\varphi(x)\psi(x)|\ dx\}\ AB + \{\int_a^b \psi^2\ dx\}B^2$$

is a positive definite from A and B. Thus, the discriminant is always non-positive:

$$\{\int_a^b |\varphi\psi|\ dx\}^2 - \int_a^b \varphi^2\ dx \cdot \int_a^b \psi^2 \quad dx \leqq 0\ .$$

The equality sign holds only if φ and ψ are <u>linearly dependent</u>.

It is noted that Schwarz' inequality is analogous to the case of the finite sums:

$$\sum_{n=1}^{m} a_n b_n \leqq \{ \sum_{n=0}^{m} a_n^2 \cdot \sum_{n=1}^{m} b_n^2\}^{\frac{1}{2}}$$

The equality holds if the set $(a_1, \ldots, a_m)$ is proportional to $(b_1, \ldots, b_m)$.

7.2 Bessel's Inequality

Let $f(x)$ be any function in the given class. Let $\varphi_1(x)$, ... $\varphi_m(x)$ be any given orthogonal normalized* set of functions in the class. We shall first show that among all the linear combinations of φ_1, ... φ_m:

$$\sum_{n=1}^{m} A_n\varphi_n ,$$

the one which makes the least error in the square mean approximation of $f(x)$, is:

$$A_n = \int_a^b f(x)\varphi_n(x)dx , \quad (n = 1,2, \dots m) . \tag{71}$$

$$K = \int_a^b \{f(x) - \sum_{n=1}^{m} C_n\varphi_n(x)\}^2 \, dx \geq 0$$

Since the set of function $\varphi(x)$ is orthonormal, we have:

$$K = \int_a^b f^2 \, dx - 2\sum_{n=1}^{m} C_nA_n + \sum_{n=1}^{m} C_n^2$$

where A_n is given by (71). On the other hand, we can also write:

$$K = \int_a^b f^2 \, dx - \sum_{n=1}^{m} A_n^2 + \sum_{n=1}^{m} (C_n-A_n)^2 .$$

Among all the choice of C_n, it is seen that K is minimum when $C_n = A_n$ $(n = 1,2, \dots m)$.

Since $K \geq 0$, we have the following important lemma:

---

*The functions φ_1, ... φ_m are called orthogonal and normalized, or orthonormal, if

$$\int_a^b \varphi_j\varphi_n \, dx = \begin{cases} 0 \text{ for } j \neq n \\ 1 \text{ for } j = n \end{cases}$$

Bessel's Inequality

$$\sum_{n=1}^{m} A_n^2 \leq \int_a^b f^2 \, dx \tag{72}$$

Corollary. If φ_n (n = 1,2,3 ...) form an infinite orthonormal set, the sum of the squares of the generalized Fourier coefficients converges; namely:

$$\sum_{n=1}^{\infty} A_n^2 \leq \int_a^b f^2(x) \, dx$$

where A_n is given by (71).

7.3 Completeness

Definition 1. A set of functions $\varphi_n(x)$ (n = 1,2,3, ...) in the class K_2 is said to be complete (vollstandig) if any function f(x) in the class can be approximated "in the sense of the square mean with an error less than any preassigned value by a linear combination of a fininte number of these functions; namely, if for any given $\varepsilon < 0$, there exists a set of constants B_n (n = 1, ... m) (some of them may be zero), such that:

$$\int_a^b \left| f(x) - \sum_{n=1}^{m} B_n \varphi_n \right|^2 dx < \varepsilon$$

If the set of functions $\varphi_n(x)$ is orthonormal, then by the remark in Section 7.2 a better approximation can be obtained by taking the generalized Fourier coefficients for the constants B_n. Furthermore, we have:

Theorem 1. An orthonormal set of functions $\varphi_n(x)$ (n = 1,2, ...) is complete if and only if the Parseval equality:

$$\sum_{n=1}^{\infty} A_n^2 = \int_a^b f^2(x) \, dx \tag{73}$$

Holds for all functions $f(x)$ in the class, where A_n are the generalized Fourier coefficients as defined by (71).

With respect to a complete orthonormal set, the generalized Fourier series always converges *in the mean* to the function; more precisely:

$$\lim_{m\to\infty} \int_a^b |f(x) - \sum_{n=1}^{m} A_n \varphi_n(x)| \ dx = 0 \ .$$

The following notation is generally used:

$$\lim_{m\to\infty} \sum_{n=1}^{m} A_n \varphi_n(x) = f(x) \ .$$

Definition 2. A set of functions is said to be *closed* (abgeschossen) in the class K_2 if the following assertion is true: If $f(x)$ is orthogonal to all functions in the set, then:

$$\int_a^b f^2(x) \ dx = 0 \ .$$

Corollary of the Theorem. *If an orthonormal set is complete, then it is closed.*

So far the integrals under consideration are either in the ordinary sense of Riemann or in the sense of Lebesgue. The converse of the above corollary can be proved for the class L_2 of functions which are square integrable in the sense of Lebesgue. The subject of Lebesgue integration is, of course, beyond the scope of the present treatment.

7.4 Example. The Completeness of Laguerre Functions in the Infinite Interval [0, ∞].

In this section the approximation is in the sense of the square mean.

Lemma 1. For any given $\varepsilon > 0$ and e^{-ax} $(a > 0)$, there is a linear combination of the orthonormal Laguerre functions:

$$S_m(x) = \sum_{n=0}^{m} A_n \varphi_n(x), \quad \text{where} \quad \varphi_n(x) = e^{-\frac{x}{2}} L_n(x)/n!$$

such that:

$$\int_0^\infty |e^{-ax} - S_m(x)|^2 \, dx < \varepsilon$$

Proof. From the generating function:

$$\frac{1}{1-t} e^{-\frac{tx}{1-t}} = e^{\frac{x}{2}} \sum_{n=0}^{\infty} t^n \varphi_n(x)$$

(c.f. Ex. 1 (iii), Section 6), it follows that:

$$g(x,t) = \frac{1}{1-t} e^{\frac{1}{2}\frac{1+t}{1-t}x} = \sum_{n=0}^{\infty} \varphi_n(x) t^n .$$

Consider:

$$\int_0^A g(x,t)\varphi_m(x) = dx = \sum_{n=0}^{N} t^n \int_0^A \varphi_m \varphi_n \, dx + R_N$$

where:

$$R_N = \sum_{n=N+1}^{\infty} t^n \int_0^A \varphi_m \varphi_n \, dx$$

For any given ε, one can choose N large enough such that:

$$N(<m) \quad \text{and} \quad |R_N| \leq \sum_{n=N+1}^{\infty} t^n = \frac{t^{N+1}}{1-t} < \varepsilon$$

for a fixed value of t ($|t| < 1$) and all values of A. Thus, when $A \to \infty$,

$$\int_0^\infty g(x,t)\, \varphi_m(x) = t^m \tag{74}$$

On the other hand,

$$\int_0 g^2(x,t) \, dx = \frac{1}{1-t^2} \tag{75}$$

From (74) and (75), it follows:

$$\int_0^\infty |g(x,t) - \sum_0^N t^n \varphi_n(x)|^2 \, dx = \frac{1}{1-t^2} - \sum_{n=0}^N t^{2n}$$

This proves the lemma.

Lemma 2. Every square integrable function f(x) in [0,1] can be approximated as close as possible in the mean by some linear combination (of a finite number) of the simple step functions:

$$\psi_{\alpha\,\beta}(x) = \begin{cases} 1 & (\alpha \leqq x \leqq \beta) \\ 0 & \text{otherwise} \end{cases}$$

Proof. We are going to give the proof for the class R_2 of Riemann square integrable functions on [0,1]. The proof for L_2 on [0,1] may be referred to in a standard text on that subject. Let us first consider the case $|f(x)| < A$. Assume that $f(x) \geqq 0$, otherwise consider $f(x) + A$ instead of f(x). From the definition of the Riemann integral, for any given ε, there is a subdivision of the interval [0,1]:

$$0 = x_1 < x_2 < \ldots < x_k = 1, \quad \text{such that:}$$

$$\left| \int_0^1 f^2 \, dx - \sigma_k \right| < \varepsilon$$

where:

$$\sigma_k(x) = \sum_{j=1}^k m_j^2 (x_j - x_{j-1})$$

$$m_j = \text{the mean value of } f(x) \text{ in } x_{j=1}, x_j$$

$$= \frac{1}{x_i - x_{i-1}} \int_{x_{i-1}}^{x_i} f(x) \, dx$$

which lies between the least upper and greatest lower bounds of f(x).

Put:

$$\psi_j(x) = \text{the simple step function on } [x_{j-1}, x_j]$$

we have:

$$\int_0^1 \psi_j^2\,dx = x_j - x_{j-1} \quad \text{by the definition.}$$

Thus:

$$\int_0^1 |f(x) - \sum_{j=1}^{k} m_j\psi_j(x)|^2\,dx$$

$$= \int_0^1 f^2\,dx - 2\sum_{j=1}^{k} m_j\int_{x_{j-1}}^{x_j} f(x)\,dx + \sum_{j=1}^{k} m_j^2(x_j-x_{j-1})$$

$$= \int_0^1 f^2\,dx - m_j^2\,(x_j-x_{j-1}) < \varepsilon$$

The proof can now be extended to the improper Riemann integral by cutting off an interval small enough in the neighborhood of points where the integrand may become infinite, so that the sum of the integrals of $f^2(x)$ over such intervals is less than any preassigned value.

Corollary. Every square integrable function on [0,1] can be approximated as close as possible by a continuous function in the mean.

The corollary is evident because of the fact that every simple step function can be approximated in the mean as close as possible by a continuous function which is represented by the broken lines.

Lemma 3. (Weierstrass) Every continuous function on [0,1] can be approximated as close as possible by a polynomial.

The proof of this lemma is usually given in the course of the advanced calculus and is omitted here. (See, for instance, Gourset-Hedrick, Mathematical Analysis, pp. 442-443, Vol. I).

Theorem*. The Laguerre functions:

$$\varphi_n = e^{-\frac{x}{2}} L_n(x)/n!$$

form an orthonormal set which is complete in the space of the square integrable functions on $[0,\infty]$.

Proof. Let $\xi = e^{-x}$ and $f(x) = F(\xi)$.
Then,

$$\int_0^\infty f^2(x)\, dx = \int_0^1 \{F^2(\xi)/\xi\}\, d\xi$$

By the Lemma 3, $F(\xi)/\sqrt{\xi}$ can be approximated in the mean as close as possible by a polynomial in ξ; namely for any given ε,

$$\int_0^1 \left| \frac{F^2(\xi)}{\xi} - \sum_{n=0}^{m} a_n \xi^n \right|^2 d\xi < \varepsilon$$

or:

$$\int_0^\infty \left| f^2(x) - \sum_{n=0}^{m} a_n\, e^{-(n+1)x} \right|^2 dx$$

By the Lemma 1, there exists A_n $(n = 0,1, \ldots p)$ such that:

$$\int_0^\infty \left| f^2(x) - \sum_{n=1}^{p} A_n \varphi_n(x) \right| dx$$

$$\leq \int_0^\infty \left| f^2(x) - \sum_{n=0}^{m} a_n\, e^{-(n+1)} x \right| dx$$

$$+ \int_0^\infty \left| \sum_{n=0}^{m} a_n\, e^{-(n+1)x} - \sum_{n=1}^{p} A_n \varphi_n(x) \right| dx \leq 2\varepsilon$$

The same arguments can be applied to the Hermite functions.

---

*The present simple proof is due to J.V. Neumann.

CHAPTER III.

THE OPERATIONAL METHOD (I) LAPLACE AND FOURIER TRANSFORMS

The methods developed in the preceding chapters fail when we deal with the boundary problems such as the type (ii) in Section 2 of the Chapter I and the heat conduction in a slab of infinite width. These problems naturally introduce the general method of operational calculus, including the integral transformations and Cauchy's residue theory. Laplace and Fourier transforms are treated in details in this chapter and their applications to the boundary value problems are illustrated by various concrete examples. Further applications of other integral transforms, such as Mellin transforms, Hankel transforms, etc., are found in the Chapter IV. In many cases, in order to find the inverse transforms, Cauchy's residue theory is indispensable. For that purpose, a brief account of the complex functions is given in the appendix. It is meant for those who do not have any previous knowledge on that subject. The theory, started from the system of complex numbers, is carried on to the Cauchy's expansion of a meromorphic function in terms of rational functions so that the students are able to apply the elementary theory of complex functions freely to the problems arisen in this course.

1. LAPLACE TRANSFORMS. ELEMENTARY RULES.

Definition. Let $F(t)$ be any function such that $F(t)\,e^{-st}$ is integrable on $[0, \infty]$ for some real value $s = s_0$. Then it is integrable for $s > s_0$.* The Laplace transform of $F(t)$ is defined as

$$L\{F(t)\} \equiv \bar{F}(s) = \int_0^\infty e^{-st}\,F(t)\,dt \qquad (1)$$

*It is shown later to be integrable for any complex quantity s such that $\mathrm{Re}(s) > s_0$, where "Re" denotes the operation of taking the real part.

for $s > s_o$.

Laplace transformations are linear (homogeneous) operators. Observe that $\bar{F}(s) \to 0$ as $s \to \infty$. In the following discussion, the Laplace transforms of the functions, whenever they appear, are assumed to exist. The following two elementary rules play the fundamental role in solving boundary value problems. It is advised that they are remembered, although their proofs are very simple.

1.1 Rule 1.* The Laplace Transform of the Derivative.

$$L\{F'(t)\} = s\bar{F}(s) - F(0) \tag{2}$$

Corollary. $L\{F^{(n)}(t)\} = s^n\bar{F}(s) - s^{n-1}F(0) - s^{n-2}F'(0) - \ldots - F^{(n-1)}(0)$ (3)

for $n = 1,2,3, \ldots$

where $F^{(n)}(t)$ is the nth derivative of $F(t)$.

Proof. It is remarked that if $e^{-st}F'(t)$ and $e^{-st}F(t)$ are integrable on $[0, \infty]$ for some values of s, then $e^{-st}F(t) \to 0$ as $t \to \infty$. This follows from the fact that if $f'(x)$ and $f(x)$ are integrable in $[0,\infty]$ then $f(x) \to 0$ as $x \to \infty$, because

$$\left|\int_A^B f'(x)\,dx\right| = |f(B) - f(A)|$$

can be made as small as possible, provided A and B are large enough, thus by Cauchy's necessary and sufficient condition for the existence of the limit as proved in advanced calculus,

$$\lim_{t\to\infty} f(x)$$

*The rule is valid provided all the terms are meaningful; for instance, the Laplace transform of both $F(t)$ and $F'(t)$ exist for a value of s.

exists. On the other hand, since f(x) is integrable, the limit must be zero. Now

$$\frac{d}{dx}[e^{-sx}f(x)] = e^{-sx}f'(x) + se^{-sx}f(x)$$

so the assertion in the beginning of the paragraph is evident. Now, integration by parts gives:

$$L(F') = \int_0^\infty F'(x)e^{-st}dt = F(x)e^{-st}\Big|_0^\infty$$

$$+ \int_0^\infty F(t)e^{-st}\,dt = s\,\bar{F}(s) - F(0)$$

The corollary follows immediately by the repeated use of (2).

1.1.1. Example. Linear Circuits. By means of Laplace transformations, the solution of the initial value problem of a system of ordinary linear differential equations such as those from the theory of linear circuits is immediately reduced to that of a system of linear algebraic equations. The inverse transforms give the solution of the differential system with the initial conditions being automatically satisfied. For instance, consider

$$y'' + y - z'' = \sin t \tag{4}$$

$$-y'' + z'' + z' + z = 2\cos t \tag{5}$$

with the initial conditions:

$$y(0) = 1, \quad y'(0) = 0 \tag{6}$$

$$z(0) = 0, \quad z'(0) = 1 \tag{7}$$

where the prime denotes the differentiation with respect to t. This system has the physical interpretation as indicated in Figure 6, where the notations are defined as in Section 5, Chapter I, except that M = the mutual inductance.

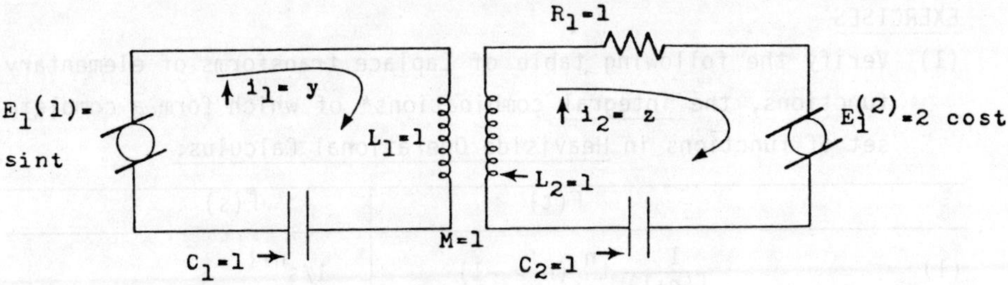

Figure 6

Let $L\{y\} = \bar{y}(s)$ and $L\{z\} = \bar{z}(s)$. It is immediately verified from the definition that

$$L\{\sin at\} = a/(s^2+a^2) \tag{8}$$

$$L\{\cos at\} = s/(s^2+a^2) \tag{9}$$

By means of (3), (6) and (7), we have

$$L\{y''\} = s^2\bar{y} - s$$

$$L\{z'\} = s\,\bar{z}$$

and

$$L\{z''\} = s^2\bar{z} - 1$$

Thus (4) and (5) become

$$(s^2+1)\,\bar{y} - s^2\bar{z} = \frac{1}{s^2+1} + s - 1$$

$$-s^2\bar{y} + (s^2+s+1)\bar{z} = \frac{2s}{s^2+1} - s + 1$$

which yield the solution

$$\bar{y} = \frac{s}{s^2+1}\,, \qquad \bar{z} = \frac{1}{s^2+1}$$

According to (8) and (9),

$$y = \cos t\,, \qquad z = \sin t\,.$$

EXERCISES

(1) Verify the following table of Laplace transforms of elementary functions, the integral combinations* of which form a complete set of functions in Heaviside Operational Calculus:

	F(t)	$\bar{F}(s)$
(1)	$\frac{1}{\Gamma(n+1)} t^n$ $(n > -1)$	$1/s^{n+1}$
Special Cases		
(1.1)	$(n = 0)$, 1	$1/s$
(1.2)	$(n = \text{integer})$, $\frac{1}{n!} t^n$	$1/s^{n+1}$
(2)	sin at	$a/(s^2+a^2)$
	cos at	$s/(s^2+a^2)$
(3)	$e^{at}F(t)$	$\bar{F}(s-a)$
Special Cases		
(3.1)	$e^{at}\, t^n/\Gamma(n+1)$	$1/(s-a)^{n+1}$
(3.2)	e^{at} sin bt	$b/\{(s-a)^2+b^2\}$
	e^{at} cos bt	$(s-a)/\{(s-a)^2+b^2\}$

The formula (2) can be considered as the corollaries of the special cases (3.1) (when $n = 0$ and ai is taken instead of a) by taking the imaginary and the real parts respectively. It is further remarked that the formula (3.1) alone is sufficient (by means of partial fractions) for finding the inverse transforms of any rational function $R(s)$ in s satisfying the necessary condition $R(s) \to 0$ as $s \to \infty$.

*An integral combination means a function obtained from the set of functions by a finite number of the following operations:
(1) Taking a linear combination; and (2) Multiplication.

(2) Let $f(t) = \int_0^t F(t)\, dt$. Show that $\bar{f}(s) = \bar{F}(s)/s$ provided both of the Laplace transforms exist at s.

(3) <u>Heaviside's Formula</u>

Let p(s) and q(s) be polynomials in s such that (i) $p(s)/q(s) \to 0$ as $s \to \infty$; and (ii) all the roots (a_1, a_2, ... a_n) of q(x) are simiple, show that the inverse Laplace transform

$$L^{-1}\left\{\frac{p(s)}{q(s)}\right\} = \sum_{j=1}^{n} \frac{p(a_j)}{q'(a_j)} e^{a_j t} \tag{10}$$

<u>HINT</u>. By partial fraction, show that

$$\frac{p(s)}{q(s)} = \sum_{j=1}^{n} \frac{A_j}{s-a_j} \tag{11}$$

where

$$A_j = \frac{p(a_j)}{q'(a_j)} \tag{12}$$

(12) is obtained by multiplying (11) by $s-a_j$ and taking the limit as $s \to a_j$.

1.2 Rule 2. Convolution.

<u>Definition.</u>

$$H(t) = \int_0^t F(t-\tau)\, G(\tau)\, d\tau$$

is defined as the <u>convolution</u> of F(t) and G(t). Sometimes it is denoted by F*G.

By a change of the variable in the integrand, $t-\tau = T$, it is readily shown that F*G = G*F.

<u>Theorem</u>.

$$\bar{H}(s) = \bar{F}(s) \cdot \bar{G}(s) \tag{13}$$

<u>Proof</u>.

$$\bar{H}(s) = \int_0^\infty e^{-st}\, dt \int_0^t F(t-\tau) G(\tau) d$$

$$= \int_0^\infty G(\tau)\, d\tau \int_0^\infty e^{-st}\, F(t-\tau)\, dt$$

$$= \int_0^\infty G(\tau)\, d\tau \int_0^\infty e^{-s(t_1+\tau)} F(t_1)\, dt_1$$

$$= \bar{G}(s) \cdot \bar{F}(s)$$

The interchange of the order of integration is legitimate under the conditions that both

$$e^{-st} F(t) \quad \text{and} \quad e^{-st} G(t)$$

are piecewise continuous* and absolutely integrable in $[0, \infty]$ for a definite value s. In many applications, these conditions cannot be justified. However, the formal procedure often gives us an explicit form of the solution which can be verified directly.

1.2.1. Example. Duhamel Formula. Consider the general equation of heat conduction

$$u_t = \lambda(u) \tag{14}$$

where

$$\lambda(u) = \frac{1}{\rho c} \left\{\frac{\partial}{\partial x} (k\, u_x) + \frac{\partial}{\partial y} (k\, u_y) + \frac{\partial}{\partial z} (k\, u_z)\right\} \tag{15}$$

(cf. (2), Ch. I). Let $\bar{u}(x,y,z,s)$ be the Laplace transform of $u(x,y,z,t)$ with respect to t. We have the formal relation

$$L\, \{\lambda(u)\} = \lambda(\bar{u}) ,$$

where L denotes the operation of taking the Laplace transformation with respect to t. Let the boundary conditions be

$$u(x,y,z,0) = 0 \tag{16}$$

$$u = 0 \text{ on } S' \quad \text{and} \quad u = F(t) \text{ on } S \tag{17}$$

where S' + S is the geometrical boundary of the solid.

---

*The condition that the functions are piecewise continuous is not necessary, if the integration is taken in a more general sense than that of Riemann.

Theorem. The solution of (14) with conditions (16) and (17) is

$$u(x,y,z,t) = \frac{\partial}{\partial t} \int_0^t F(t-\tau)\, V(x,y,z,\tau)\, d\tau \qquad (18)$$

where $V(x,y,z,t)$ is the solution of the same problem when $F(t) = 1$.

Proof. The Laplace transform of equation (14) is

$$s\bar{u} = \lambda(\bar{u})$$

with the conditions

$$\bar{u} = 0 \text{ on } S' \quad \text{and} \quad \bar{u} = \bar{F}(s) \text{ on } S.$$

$\bar{V}$ satisfies the same equation with the conditions

$$\bar{V} = 0 \text{ on } S' \text{ and } \bar{V} = \frac{1}{s} \text{ on } S$$

Hence

$$\bar{u} = s\, \bar{F}(s)\, \bar{V}$$

Let $L^{-1}(s\bar{V}) = V_t(x,y,z,t)$.

By the convolution formula (13), it follows

$$u(x,y,z,t) = \int_0^t F(\tau)\, V_t(x,y,z,t-\tau)\, d\tau \qquad (19)$$

$$= \frac{\partial}{\partial t} \int_0^t F(\tau)\, V(x,y,z,t-\tau)\, d\tau \qquad (19')$$

These are alternative forms of solution (18).

EXERCISES

(1) Show that Duhamel's formula is valid, if, instead of $u = 0$ on S', it is prescribed that

$$Au + B\frac{\partial u}{\partial n} = 0 \text{ on } S'$$

where A and B are two constants and $\frac{\partial u}{\partial n}$ is the directional derivative along the normal.

(2) Find the solution for the slab:

$$u_t = u_{xx}$$

$$u(x,0) = 0$$

$$u(0,t) = 0$$

$$u(1,t) = F(t)$$

<u>HINT</u>. Set $V(x,t) = x + W(x,t)$. Then $W_t = W_{xx}$ with the conditions

$$W(x,0) = -x$$

$$W(0,t) = W(1,t) = 0$$

So

$$W(x,t) = 2 \sum_{n=1}^{\infty} \frac{(-1)^n}{n\pi} e^{-n^2\pi^2 t} \sin n\pi x .$$

Apply the Duhamel's formula (19).

<u>ANSWER</u>.

$$u(x,t) = 2 \sum_{n=1}^{\infty} (-1)^{n+1} n\pi \sin n\pi x \int_0^t F(t-\tau)\, e^{-n^2\pi^2 t}\, d\tau$$

(3) <u>Other Elementary Rules</u>.

Verify the formulas in the following table:

	$F(t)$	$\bar{F}(s)$
(3.1)	$-t\,F(t)$	$\bar{F}'(s)$
(3.2)	$F(t-a)$, $a > 0$	$e^{-as}\,\bar{F}(s)$
(3.3)	$aF(at)$, $a > 0$	$F(s/a)$

where

$$[F(t-a)] = \begin{cases} F(t-a) & \text{if} \quad t \geq a \\ 0 & \text{if} \quad t < a \end{cases}$$

The function $[F(t-a)]$ can be interpreted as to represent a delayed action. Compare the curves representing respectively

$$y = F(t) \quad \text{and} \quad y = [F(t-a)] .$$

(4) Find the inverse Laplace transform of

$$\bar{F}(s)/s^n \quad (n = \text{a positive integer})$$

by means of

(a) the convolution formula (13); and

(b) the formula 3.1 in Ex. 3.

Show that the nth repeated integral of F(t) is

$$\frac{1}{(n-1)!}\int_0^t (t-\tau)^{n-1} F(\tau)\, d\tau .$$

(The uniqueness of the inverse transform is assumed.)

(5) By means of Ex. (3.1), find the Laplace transform of the equation

$$t\frac{d^2y}{dt^2} + \frac{dy}{dt} + ty = 0$$

Such that y(0) = 1 and the derivatives of y are finite at t=0. Thus find the solution in power series.

HINT.

$$\frac{d\bar{y}}{ds} = \frac{-s}{s^2+1}\,\bar{y}$$

Hence

$$\bar{y} = c\sum_{n=0}^{\infty}(-1)^n \frac{1}{2}\,\frac{3}{2}\,\frac{2n-1}{2}\,\frac{1}{n!}\,\frac{1}{s^{2n+1}}$$

where c is arbitrary. Find the inverse transform termwise and determine c by the condition y(0) = 1.

ANSWER.

$$y = J_0(t) .$$

2. LAPLACE TRANSFORMS OF SPECIAL FUNCTIONS

Very often the difficulty of solving a boundary value problem is reduced to find inverse Laplace transform. For the future use, we calculate the Laplace transforms of some special functions.

2.1 Periodic Functions.

Theorem. If $f(t+a) = f(t)$, $a > 0$, then

$$\bar{f}(s) = \int_0^a e^{-st} f(t)\, dt/(1-e^{-as}) \tag{20}$$

Proof. From the definition,

$$\bar{f}(s) = \sum_{n=0}^{\infty} \int_{na}^{(n+1)a} e^{-st} f(t)\, dt .$$

Substituting t-na for t and using the property of periodicity, one derives

$$f(s) = \left(\sum_{n=0}^{\infty} e^{-ns} \right) \int_0^a e^{-st} f(t)\, dt$$

Since

$$\sum_{n=0}^{\infty} e^{-nas} = 1/(1-e^{-as}) ,$$

(20) follows immediately.

EXERCISE

Let

$$M_c(t) = \begin{cases} 1 , & 0 < t \leqq c \\ -1 , & c < t \leqq 2c \end{cases}$$

Define $M_c(t+2c) = M_c(t)$.
Show that

$$L\{M_c(t)\} = \frac{1}{s} \tan h \frac{cs}{2}$$

Therefore by Ex. (3.1) in the last section,

$$L\{H_c(t)\} = \frac{1}{s^2} \tan h \frac{cs}{2}$$

where

$$H_c(t) = \int_0^t M_c(t)\, dt.$$

(Observe that $H_c(t)$ is still periodic of the period 2c.)

2.2 Some Special Laplace Transforms

Consider

$$F(t) = t^{-3/2}\, e^{-k^2/4t}$$

where k is a constant. In order to find F(s), the value of the definite integral is assumed to be known.*

$$\int_0^\infty e^{-x^2}\, dx = \sqrt{\pi}/2 . \qquad (21)$$

$$\bar{F}(s) = \int_0^t t^{-3/2}\, e^{-st-k^2/4t}\, dt = \frac{4}{k}\int_0^\infty \exp(-\tau^2 - \frac{k^2 s}{4\tau^2})\, d\tau \qquad (22)$$

which is obtained by the substitution $t^{\frac{1}{2}} = k/2\tau$.

Put

$$a^2 = k^2\, s/4 .$$

Consider

$$I^* = \int_0^\infty \exp\,(-\tau^2 - \frac{a^2}{\tau^2})\, d\tau$$

$$= e^{-2a}\int_0^\infty \exp\,\{-(\tau - \frac{a}{\tau})^2\}\, d\tau$$

$$= e^{-2a}\int_0^\infty \exp\,\{-(\tau_1 - \frac{a}{\tau_1})^2\}\, \frac{a d\tau_1}{\tau_1^2}$$

The last integral is derived by the substitution $\tau_1 = \frac{a}{\tau}$. By adding the last two integrals we have

---

*Let the value of the integral be I.

$$I^2 = \int_0^\infty e^{-x^2}\, dx \int_0^\infty e^{-y^2}\, dy = \int_0^{\frac{\pi}{2}} d\theta \int_0^\infty e^{-\rho^2}\, \rho\, d\rho = \frac{\pi}{4}$$

where ρ, θ are the polar coordinates of (x,y).

$$2I^* = e^{-2a} \int_0^\infty (1 + \frac{a}{t^2}) \exp \{-(t- \frac{a}{t})^2\} \, dt$$

$$= e^{-2a} \int_{-\infty}^{\infty} e^{-x^2} \, dx = \sqrt{\pi} \, e^{-2a}$$

Substituting in (22), it follows that

$$L^{-1} \{e^{-k\sqrt{s}}\} = \frac{k}{2\sqrt{\pi} \, t^{3/2}} e^{-k^2/4t} \tag{23}$$

where L^{-1} denotes the inverse Laplace transformation.

EXERCISES

(1) Show by Ex. (3.1) of the preceding section, that

$$L \{\frac{1}{\sqrt{\pi}} e^{-k^2/4t}\} = \frac{1}{\sqrt{s}} e^{-k\sqrt{s}} \tag{24}$$

(2) Show, by the colvolution formula (13), that

$$L \{\operatorname{erfc} (\frac{k}{2\sqrt{t}})\} = \frac{1}{s} e^{-k\sqrt{s}} \tag{25}$$

where

$$\operatorname{erf}(x) = \frac{2}{\sqrt{\pi}} \int_0^x e^{-x^2} \, dx \tag{26}$$

$$\operatorname{erfc}(x) = 1 - \operatorname{erf}(x) \tag{27}$$

3. APPLICATIONS TO BOUNDARY VALUE PROBLEMS

3.1 Heat Conduction in an Infinite Wire

When a wire is of infinite length, even for the initial value problem, the method of Fourier Series cannot be applied. For instance, consider

$$u_t = u_{xx}$$

with the boundary conditions:

$$u(x,+0) = 1 \quad (0 < x)$$

$$u(0+,t) = 0 \quad (0 < t)$$

and

$$\lim_{x\to\infty} u(x,t) \quad \text{is bounded.}$$

In order to find the explicit form of the solution, the method of Laplace transformation is employed. The operation is performed formally. Not every step can be justified, nor is it necessary to do so in many instances, as far as the solution of the problem is concerned. The solution, once obtained explicitly, can then be verified directly. Let $L\{u\}$ be the Laplace transform of $u(x,t)$ with respect to t and denoted by $\bar{u}(x,s)$. One has, formally

$$L\{u_x\} = \bar{u}_x$$

The heat equation is transformed into

$$s\bar{u} - 1 = \bar{u}_{xx} \tag{28}$$

The boundary conditions are:

$$\bar{u}(0,s) = 0$$

and

$$\bar{u}(x,s) \quad \text{is bounded.}$$

For each s, the equation (28) can be considered as an ordinary differential equation. The solution satisfying the boundary conditions is

$$\bar{u}(x,s) = \frac{1}{s}\,(1-e^{-x\sqrt{s}})$$

By the exercise (2) in the preceding section

$$u(x,t) = \operatorname{erf}\frac{x}{2\sqrt{t}} = \frac{2}{\sqrt{\pi}}\int_0^{\frac{x}{2\sqrt{t}}} e^{-\xi^2}\, d\xi\,. \tag{29}$$

To verify that this is a solution of the heat equation satisfying the boundary conditions is then a simple affair.

EXERCISES

(1) Find the solution of the equation

$$u_{tt} = u_{xx}$$

with the boundary condition

$$u(x,0) = 0 \ , \quad u_t(x,0) = 1 \qquad (x < 0)$$

$$u(0,t) = 0 \ , \quad \lim_{x\to\infty} u_x(x,t) = 0 \quad (t > 0) \ .$$

The problem can be interpreted as that an elastic bar is moving with unit velocity when suddenly one end is clamped to rest while the other end remains free.

ANSWER.

$$u(x,t) = \begin{cases} t & (t \leqq x) \\ x & (t \geqq x) \end{cases}$$

(2) Find the longitudinal displacement of an elastic bar which is in longitudinal vibration due to a prescribed constant force acting on one end, while the other end is kept fixed,

$$u_{tt} = u_{xx} \qquad (0 < x < c, \quad t > 0)$$

$$u(x,0) = u_t(x,0) = 0 \qquad (0 < x < c)$$

$$u(0,t) = 0 \ , \quad u_x(c,t) = F. = \text{Constant} \quad (t < 0)$$

where both the velocity of sound and Young's modulus are normalized to unity. Show, by the formula in the exercise of Section 2.1, that the longitudinal displacement measured from its initial position at a point x is indicated as in the following figure.

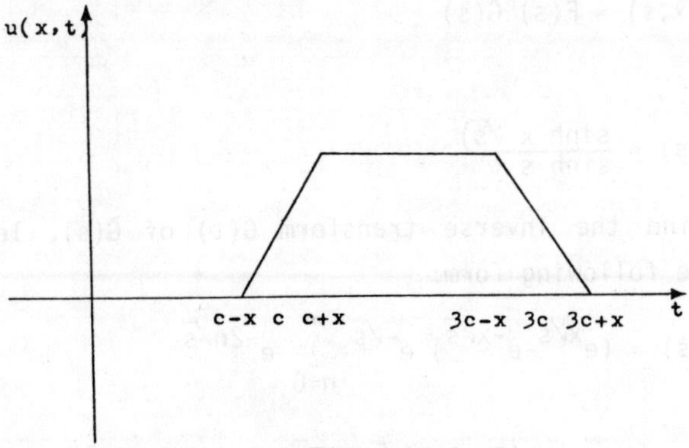

Figure 7

(c is a fixed parameter.)

ANSWER.

$$\bar{u}(x,s) = F \cdot \frac{1}{s^2} \frac{\sinh (xs)}{\cosh (cs)}$$

$$= F \cdot \left[\frac{1}{s^2} + \frac{1}{s^2} \tanh (cs)\right] \left[e^{-(c-x)s} - e^{-(c+x)s}\right]$$

3.2 The Heat Conduction in a Slab.

The solution of the heat conduction in a slab is given by Duhamel's formula in the exercise (2) of Section 1.2.1. It is easily seen to be convergent for $t > 0$. However, it converges slowly for small values of t. Another form of the solution, which converges quickly for small values of t can be obtained by finding the inverse transform of the solution in a different way.

Consider the partial differential equation

$$u_t = u_{xx} \qquad (0 < x < 1 , \quad 0 < t) \tag{30}$$

with the boundary condition

$$\begin{aligned} u(x,+0) &= 0 && 0 < x < 1 \\ u(+0,t) &= 0 && 0 < t \\ u(-1,t) &= F(t) && 0 < t \end{aligned} \tag{31}$$

we have,

$$\bar{u}(x,s) = \bar{F}(s)\ \bar{G}(s) \tag{32}$$

where

$$\bar{G}(s) = \frac{\sinh x\ \sqrt{s})}{\sinh s}$$

In order to find the inverse transform $G(t)$ of $\bar{G}(s)$, let it be rewritten in the following form:

$$\bar{G}(s) = (e^{x\sqrt{s}} - e^{-x\sqrt{s}})\ e^{-\sqrt{s}} \sum_{n=0}^{\infty} e^{-2n\sqrt{s}}$$

$$= \sum_{n=0}^{\infty} e^{-(2n+1-x)\sqrt{s}} - \sum_{n=0}^{\infty} e^{-(2n+1+x)\sqrt{s}}$$

By means of the formula (23), one has

$$g(t) = \frac{1}{2\sqrt{\pi}t^{3/2}} \Big[\sum_{n=0}^{\infty} e^{-(2n+1-x)^2/4t}$$

$$- \sum_{n=0}^{\infty} (2n+1+x)\ e^{-(2n+1-x)^2/4t} \Big] \tag{33}$$

Each series of (33) is seen to be convergent uniformly for $t \geqq t_0 > 0$ and $0 \leqq x \leqq 1$. It is evident that they converge quickly for small values of t.

Now, by virtue of the convolution formula (13) and by a simple change of the variable of integration, one derives

$$u(x,t) = \frac{2}{\sqrt{\pi}} \sum_{n=0}^{\infty} \int_{\frac{2n+1-x}{2\sqrt{t}}}^{\infty} F\left(t - \frac{(2n+1-x)^2}{4\lambda^2}\right) e^{-\lambda^2}\ d\lambda$$

$$- \frac{2}{\sqrt{\pi}} \sum_{n=0}^{\infty} \int_{\frac{2n+1+x}{2\sqrt{t}}}^{\infty} F\left(t - \frac{(2n+1+x)^2}{4\lambda^2}\right) e^{-\lambda^2}\ d\lambda \tag{34}$$

3.2.1 Rigorous establishment of the solution.* Let F(t) be continuous in any finite time-interval. One finds that

$$\text{(i)}\quad \lim_{x\to+0} u(x,t) = 0$$

for a fixed value of $t > 0$. Let

$$|F(t_1)| < M(t) \quad \text{for} \quad 0 \leq t_1 \leq t$$

Then each term in either series of (34) is in absolute value less than

$$M \int_{n/\sqrt{t}}^{\infty} e^{-\lambda^2}\, d\lambda < M \int_{n/\sqrt{t}}^{\infty} e^{-{}^2}\, d\lambda = \frac{M}{2} e^{-n^2/t} = \frac{M}{2} e^{-n^2/t_1}$$

when $n^2 \geqq t$. On the other hand, it can be readily shown that each integral

$$\int_{\frac{a-x}{2\sqrt{t}}}^{\infty} F(t - \frac{(a-x)^2}{4\lambda^2})\, e^{-\lambda^2}\, d\lambda\,, \quad \text{where a is an integer,}$$

is continuous in x and t for $0 \leqq x \leqq 1$ and $0 < t_0 \leqq t \leqq t_1$. It is also continuous in the region $0 < x_0 \leqq x \leqq x_1 < 1$ and $0 \leqq t$ if the integral is defined as 0 when $t = 0$. Thus, by (35), (34) follows immediately by taking the limit term by term. Similarly, we have

$$\text{(ii)}\quad \lim_{x\to 1-0} u(x,t) = \frac{2}{\sqrt{\pi}} \int_0^{\infty} F(t)\, e^{-\lambda 2}\, d\lambda = F(t)$$

---

*The existence theorem to a mathematician is to find the weakest conditions for which a solution is rigorously found. For a practical problem, however, the realistic conditions are given, which are practically feasible, so that the solution is shown to be mathematically established. The same remark also goes for the uniqueness. In the practical problems, especially in the nonlinear case, an approximate solution is sometimes useful and meaningful only if, under the prescribed (say, boundary) conditions, a unique solution is shown or assumed to exist.

uniformly in $t(0 < t_0 \leqq t \leqq t_1)$, and

$$\text{(iii)} \quad \lim_{t \to +0} u(x,t) = 0 \quad \text{uniformly in} \quad x(0 < x_0 \leqq x \leqq x_1).$$

Moreover, we have manifestly shown that $u(x,t)$ is bounded in the region $0 \leqq t \leqq t_1$ and $0 \leqq x \leqq 1$ since each integral in (i) is bounded.

By use of the symmetrical form of the convolution, the differentiated series converges uniformly for $0 < t_0 \leqq t \leqq t_1$ and $0 \leqq x \leqq 1$. Thus it satisfies the partial differential equation (30).

3.2.2 Uniqueness. Let $F(t)$ be continuous in any finite interval $0 \leqq t \leqq t_1$.

Theorem. There exists one and only one solution of (30) satisfying the boundary conditions (31)* such that

(i) $u(x,t)$ is bounded in the region $0 \leqq x \leqq 1$ and $0 \leqq t \leqq t_1$.

(ii) $u(x,t)$ and its first derivative with respect to t are continuous in x and t for $0 < t_0 \leqq t$.

Proof. Suppose there is another solution $v(x,t)$ satisfying the same conditions. Consider $w(x,t) = u(x,t) - v(x,t)$. Then $w(x,t)$ is a solution of (30) satisfying all the homogeneous boundary conditions and also the conditions (i) and (ii).

Now, put

$$K(t) = \int_0^1 w^2(x,t)\, dx \tag{36}$$

For $t > 0$,

$$K'(t) = 2\int_0^1 w\, w_t\, dx = \int_0^1 w\, w_{xx}\, dx$$

*The conditions can of course be weakened. It is also remarked that the question of uniqueness for general case of non-homogeneous boundary conditions can be treated similarly.

$$= w_x w \Big|_0^1 - \int_0^1 w_x^2 \, dx = -\int_0^1 w_x^2 \, dx \leqq 0 .$$

All these operations are justified because of (ii). Therefore $K(t) \geqq 0$ is monotonically decreasing function of t. Thus K(0+) exists and is equal to*

$$\int_0^1 w^2(x,+0) \, dx = 0 \text{ , by (ii) and } w(x,+0) = 0.$$

Thus $K(t) \equiv 0$. From (36) one derives immediately $w(x,t) = 0$ for $0 \leqq x \leqq 0$, and $t > 0$. This completes the proof of the theorem.

It is worthy of note that the solution V(x,t), obtained in Ex. (2) of Section 1.2.1 by Duhamel's formula can be seen to satisfy all the conditions of the theorem; therefore, the solution should be the same as (34). The verification is left to the students as an exercise.

3.3 Transients in a Semi-Infinite Transmission Line

Consider the equations for a transmission line of 5, Chapter I:

$$v_x + Li_t + Ri = 0 \tag{37}$$

$$Cv_t + Gv + i_x = 0 \tag{38}$$

with the <u>quiescent</u> initial conditions

$$v(x,+0) = i \;\; (x,+0) = 0 \tag{39}$$

and the end conditions

$$v(+0,t) = v_o(t) \text{ and } \lim_{t\to\infty} v(x,t) = 0 \tag{40}$$

---

*The assertion is true if one is allowed to use Lebesque theory to justify it. If, however, one is limited to the knowledge of Riemann theory, he may make use of (iii) of the last section as an additional assumption.

or

$$i(+0,t) = i_0(t) \quad \text{and} \quad \lim_{x\to\infty} i(x,t) = 0 \tag{41}$$

The Laplace transform with respect to t of the solution is

$$\bar{v}(x,s) = \bar{v}_0(s)\,\bar{g}(x,s) \tag{42}$$

where

$$\bar{g}(x,s) = \exp\left(-x\sqrt{(Ls+R)(Cs+G)}\right) \tag{43}$$

A similar form is obtained for i(x,s) if the condition (41) is taken into account instead of (40). Then

$$v(x,t) = v_0(t) * g(x,t)$$

It should be remarked that the corresponding i(x,t) is found by solving (37) with the condition (39). The problem of finding g(t) in general is not a simple matter without the use of the <u>inversion formula</u> (cf. Section 4), which requires the Cauchy residue theory (as given in the appendix). However, some important special cases can be readily solved by the virtue of the elementary rules (cf. the exercises following this section).

<u>EXERCISES</u>

Find the solutions for the following special case of the semi-infinite transmission lines:

(1) Signaling through a leakage free, non-inductive cable, L = G = 0.

<u>ANSWER</u>. $$v(x,t) = \frac{x\sqrt{RC}}{2\sqrt{\tau}} \int_0^t \frac{e^{-\frac{x^2 RC}{4\tau}}\, V_0(t-\tau)}{\tau^{3/2}}\, d\tau$$

(2) Distortionless transmission.

$$\frac{R}{L} = \frac{G}{C}$$

<u>ANSWER</u>.

$$v(x,t) = e^{-\frac{R}{L} x \sqrt{LC}} [v_o(t-x\sqrt{LC})]$$

where the bracket is defined as in Ex. (3) in Section 1.2.1.

4. THE COMPLEX INVERSION FORMULA

The general problem of finding the inverse Laplace transform requires further considerations. The following formula is found to be particularly useful in many applications:

$$F(t) = L^{-1} \{\bar{F}(s)\} = \frac{1}{2\pi i}\int_{\gamma-\infty i}^{\gamma+\infty i} \bar{F}(s)\, e^{ts}\, ds \qquad (44)$$

where γ is a positive constant (chosen large enough) and the value of the integral is independent of γ as shown later. If one is not familiar with the theory of complex integration and Cauchy's residue theory, the appendix should be consulted first before going further. As it is remarked in a footnote in the beginning of this chapter, if F(s) exists for $s = s_0$, then F(s) is an analytic function for s in the right half-plane $Re(s) > s_0$, where Re(s) denotes the real part of s. The advantage of (44) is that the integral can often be evaluated by means of the Cauchy residue theory. The precise assertion concerning the validity of such a formula and its proof are postponed until the inversion formula is established for Fourier transforms (see Section 5).

4.1 Example

(1) Heaviside's Formula.

Let p(s) and q(s) be polynomials as defined in Ex. (3) of Section 1.1.1. Consider the complex integration.

$$\frac{1}{2\pi i}\oint_B \frac{p(s)}{q(s)}\, e^{ts}\, ds = \sum_{j=1}^{n} \text{residue at } a_j$$

where $B = B_1+B_2$ is the contour as indicated in the figure.

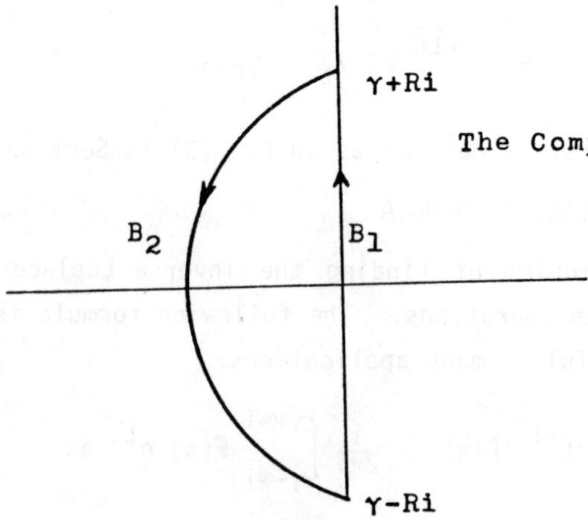

Figure 8

As $R \to \infty$, the integral along $B_2 \to 0$ while that along B_1 gives the required integral for the inverse transform (44). On the other hand, the residue of the integrand of $s = a_j$ is

$$\lim_{s \to a_j} e^{ts} p(s) \frac{s-a_j}{q(s)} = e^{a_j t} p(a_j)/q'(a_j)$$

Thus the Heaviside formula follows.

(2) The inverse transform of $\bar{f}(s) = \frac{1}{\sqrt{s^2-\beta^2}} e^{-(s^2-\beta^2)^{\frac{1}{2}}}$

By (44),

$$f(t) = \frac{1}{2\pi i} \int_{\gamma-i\infty}^{\gamma+i\infty} \frac{1}{\sqrt{s^2-\beta^2}} e^{-(s^2-\beta^2)^{\frac{1}{2}}+st} ds \qquad (45)$$

In order to rationalize the integrand, put

$$\xi^{\frac{1}{2}} = (s+\beta)^{\frac{1}{2}} + (s-\beta)^{\frac{1}{2}} .$$

Then

$$2\beta\xi^{-\frac{1}{2}} = (s+\beta)^{\frac{1}{2}} - (s-\beta)^{\frac{1}{2}}$$

$$s = \frac{1}{4} (\xi + \frac{4\beta^2}{\xi}) ,$$

$$(s^2-\beta^2)^{\frac{1}{2}} = \frac{1}{4}\,(\xi - \frac{4\beta^2}{\xi}),$$

and

$$\frac{d\xi}{\xi} = \frac{ds}{(s^2-\beta^2)^{\frac{1}{2}}}$$

Therefore

$$f(t) = \frac{1}{2\pi i}\int_{\gamma'-\infty i}^{\gamma'+\infty i} \{e^{\frac{1}{4}(t-1)\xi+\beta^2(t+1)/\xi}\}\ \frac{d\xi}{\xi} \qquad (46)$$

where the transformed contour on the ξ-plane is deformed into a straight line by Cauchy's fundamental theorem.

Now one has to distinguish between the following two cases:

(i) $t < 1$. It is immediately seen, again by Cauchy's theorem, that the integral in (46) is equal to the limit of a contour integration along a semi-circle in the right half plane joining the points $\gamma'+Ri$ to $\gamma'-Ri$. The latter tends to zero as $R \to 0$. Therefore

$$f(t) = 0 \quad \text{for} \quad t < 1. \qquad (47)$$

(ii) $t > 1$. By the virtue of the substitution

$$\frac{1}{4}\,(t-1)\xi = \zeta$$

$$f(t) = \frac{1}{2\pi i}\int_{\gamma''-i\infty}^{\gamma''+i\infty} \{e^{\zeta+\sigma^2/4\zeta}\}\frac{d\zeta}{\zeta} \qquad (48)$$

where $\beta^2 = \sigma^2\ (t^2-1)$. It can be shown that the integral (48) is the Bessel function of order zero with pure imaginary argument $J_0(i\sigma) = I_0(\sigma)$. As a matter of fact, it is true that

$$\frac{1}{2\pi i}\int_{\gamma-i\infty}^{\gamma+i\infty} [e^{\zeta+\sigma^2/4\zeta}]\ \frac{d\zeta}{\zeta^{n+1}} = I_n(\sigma)\cdot(\frac{2}{\sigma})^n \qquad (49)$$

where $n > -1$. Because

$$I_n(\sigma) = \sum_{j=0}^{\infty} \frac{1}{j!\Gamma(n+j+1)} \left(\frac{\sigma}{2}\right)^{2j+n}$$

$$= \sum_{j=0}^{\infty} \frac{1}{j!} \left(\frac{\sigma}{2}\right)^{2j+n} \frac{1}{2\pi i} \int_{\gamma-i\infty}^{\gamma+i\infty} \frac{e^{\zeta} d\zeta}{\zeta^{n+j+1}}$$

$$= \frac{1}{2\pi i} \left(\frac{1}{2}\sigma\right)^n \int_{\gamma-i\infty}^{\gamma+i\infty} \frac{e^{\zeta}}{\zeta^{n+1}} \sum_{j=0}^{\infty} \frac{1}{j!} \left(\frac{\sigma^2}{4\zeta}\right)^j d\zeta$$

$$= \frac{1}{2\pi i} \left(\frac{\sigma}{2}\right)^n \int_{\gamma-i\infty}^{\gamma+i\infty} e^{\zeta+\sigma^2/4\zeta} \frac{d\zeta}{\zeta^{n+1}}$$

The second equality is obtained by the fact that

$$L^{-1}\left\{\frac{1}{s^{k+1}}\right\} = \frac{1}{\Gamma(k+1)} t^k , \quad (k> -1) .$$

Then set $t = 1$. The interchange of the integral sign and the summation sign can also be justified.

From (i) and (ii), one derives

$$L^{-1}\left\{\frac{1}{\sqrt{s^2=\beta^2}} e^{-(s^2-\beta^2)^{\frac{1}{2}}}\right\} = I_0 \left(\beta \sqrt{t^2-1}\right) E(t-1) \qquad (50)$$

where $E(t)$ is the Heaviside unit step function, defined in Section 7.2, Chapter I.

It is now a simple matter to find the function $g(t)$ in the last section by the elementary rules. Let

$$(Ls+R)(Cs+G) = \frac{1}{\alpha^2}\left\{(s+\rho)^2 - \lambda^2\right\}$$

where

$$\alpha = \frac{1}{\sqrt{LC}} , \quad \rho = \frac{R}{2L} + \frac{G}{2C} \quad \text{and} \quad \lambda = \frac{R}{2L} - \frac{G}{2C}$$

$$g(t) = L^{-1}\left\{e^{-x\sqrt{(Ls+R)(Cs+G)}}\right\}$$

$$= \frac{\partial}{\partial x} L^{-1}\left\{\frac{-1}{\sqrt{(Ls+R)(Cs+G)}} e^{-x\sqrt{(Ls+R)(Cs+G)}}\right\}$$

A simple calculation leads to

$$g(t) = L^{-}\{e^{-x\sqrt{(Ls+R)(Cs+G)}}\}$$

$$= e^{-\rho t}\,\delta(t - \frac{x}{\alpha})$$

$$+ e^{\rho t}\frac{\lambda x}{\alpha\sqrt{t^2-(\frac{x}{\alpha})^2}}\, I_1(\lambda\sqrt{t^2-(\frac{x}{\alpha})^2})\; E(t - \frac{x}{\alpha}) \qquad (51)$$

EXERCISE

Show that the corresponding solution of the induced current of the last section is

$$\bar{i}(x,s) = \bar{v}_0(s)(\frac{Cs+G}{Ls+R})^{\frac{1}{2}}\; e^{-x\sqrt{(Ls+R)(Cs+G)}}$$

Show also, for $t > \frac{x}{\alpha}$, that

$$i(x,t) = \sqrt{\frac{C}{L}}\, v_0(t - \frac{x}{\alpha})\, e^{-\frac{\rho x}{\alpha}}$$

$$+ \int_{x/\alpha}^{t} v_0(t-\tau)\, e^{-\rho\tau}\{\frac{\lambda\tau I_1(\lambda\sqrt{\tau^2-(\frac{x}{\alpha})^2})}{(\tau^2-(\frac{x}{\alpha})^2)^{\frac{1}{2}}} - \lambda I_0(\lambda\sqrt{\tau^2-(\frac{x}{\alpha})^2})\}d\tau$$

and $i(x,t) = 0$ for $t < (\frac{x}{\alpha})$.

HINT.

$$\frac{-2}{x}\frac{\partial}{\partial L}\{\frac{1}{s}\, e^{-x\sqrt{(Ls+R)(Cs+G)}}\} = (\frac{Cs+G}{Ls+R})^{\frac{1}{2}}\; e^{-x\sqrt{(Ls+R)(Cs+G)}}$$

4.2 The Viscous Flow Between Two Infinite Concentric Cylinders

As an application of Cauchy's residue theory for finding the inverse Laplace transform of a complicated function of s, consider the equation for viscous flow between two infinite concentric cylinders with radii a and b:

$$\frac{\partial^2 v}{\partial r^2} + \frac{1}{r}\frac{\partial v}{\partial r} - \frac{v}{r^2} = \frac{1}{\nu}\frac{\partial v}{\partial t} \qquad \begin{matrix} a < r < b \\ t > 0 \end{matrix} \qquad (53)$$

where r = the radial distance

t = the time

ν = the kinematic viscosity

v = the tangential velocity component normal to both the axis of the cylinders and the r-direction.

If at $t = 0$, the outer cylinder starts rotating with a constant angular velocity ω, the boundary conditions are:

$$\begin{aligned} v(r,+0) &= 0 \\ v(a+0,t) &= 0 \\ v(b-0,t) &= b\omega \end{aligned} \tag{54}$$

The Laplace transform for the solution of (53) with the conditions (54) is readily obtained as

$$\bar{v}(s) = \frac{b\omega}{s} \frac{\begin{vmatrix} I_1(a\sqrt{\frac{s}{\nu}}) & I_1(r\sqrt{\frac{s}{\nu}}) \\ K_1(a\sqrt{\frac{s}{\nu}}) & K_1(r\sqrt{\frac{s}{\nu}}) \end{vmatrix}}{\begin{vmatrix} I_1(a\sqrt{\frac{s}{\nu}}), & I_1(b\sqrt{\frac{s}{\nu}}) \\ K_1(a\sqrt{\frac{s}{\nu}}) & K_1(b\sqrt{\frac{s}{\nu}}) \end{vmatrix}} \tag{55}$$

where

$$I_1(is) = i\, J_1(s), \quad K_1(is) = -\frac{1}{2}\pi\{J_1(s) - i\, Y_1(s)\} \tag{56}$$

From the definition of Bessel functions, it is readily shown that $v(x,s)$ is single-valued in s, in the whole plane. Put $s = -\nu\alpha^2$. The zeros of the denominator are $\alpha = 0$ and the roots of

$$\begin{aligned} &I_1(ia\alpha)K_1(ib\alpha) - K_1(ia\alpha)\, I_1(ib\alpha) \\ &\quad = -\frac{\pi}{2}\{J_1(a\alpha)Y_1(b\alpha) - Y_1(a\alpha)J_1(b\alpha) = 0\ . \end{aligned} \tag{57}$$

The expression on the left side, being considered as a function of α, is a solution of a Sturm-Liouville system with a parameter α^2. Hence all the roots (57) are real and are positive and negative in pairs. Let α_1, α_2, ... α_n ... be all the positive roots of (57).

Set

$$k_j = \frac{J_1(a\alpha_j)}{J_1(b\alpha_j)} = \frac{Y_1(a\alpha_j)}{Y_1(b\alpha_j)}$$

Now since the Wronskian

$$\begin{vmatrix} J_1(s) & Y_1(s) \\ J_1'(s) & Y_1'(s) \end{vmatrix} = \frac{2}{\pi s},$$

one derives

$$\frac{d}{d\alpha}\begin{vmatrix} J_1(a\alpha) & Y_1(a\alpha) \\ J_1(b\alpha) & Y_1(b\alpha) \end{vmatrix}_{\alpha=\alpha_j} = \frac{2}{\pi\alpha_j}\left(k_j - \frac{1}{k_j}\right) = \frac{2}{\pi\alpha_j}\,\frac{J_1^2(a\alpha_j)-J_1^2(b\alpha_j)}{J_1(a\alpha_j)J_1(b\alpha_j)}$$

which is non-vanishing as the roots of (57) can be shown to be simple.* Therefore the residue of $\frac{\bar{v}(x,s)e^{st}}{b\omega}$ at $s = -\nu\alpha_j^2$ is

$$R_j = \frac{J_1(a\alpha_j)J_1(b\alpha_j)}{J_1^2(a\alpha_j)-J_1^2(b\alpha_j)}\begin{vmatrix} J_1(a\alpha_j) & Y_1(r\alpha_j) \\ J_1(b\alpha_j) & Y_1(r\alpha_j) \end{vmatrix} e^{-\nu\alpha_j^2 t} \qquad (58)$$

On the other hand the residue at $s = 0$ is

$$\frac{b}{r}\,\frac{r^2-a^2}{b^2-a^2}$$

Now the formula (44) for $\bar{v}(x,s)$ can be evaluated formally by the residue at all the poles of $\bar{v}(x,s)$ as in the example (1) of the last section. Thus, one derives

---

*See, for instance, Gray and Mathews, Treatise on Bessel Functions (1922), p. 82. It can also be seen that $s = \nu\alpha^2$ are the eigenvalues of a Sturm-Liouville system. Since the differential equation is of second order, corresponding to each eigenvalue, the eigenfunction is uniquely determined up to a constant factor. Thus, it must be a simple zero.

$$v(r,t) = \frac{b^2}{r}\frac{r^2-a^2}{b^2-a^2} - b\omega \sum_{j=0}^{\infty} R_j \tag{59}$$

where R_j is given by (58).

<u>EXERCISE.</u>

<u>Falling of a Heavy Thin Lamina Through a Viscous Fluid.</u>

Find the solution of the equation

$$\frac{\partial v}{\partial t} = \nu \frac{\partial^2 v}{\partial x^2} \quad (0 < |x| < 1 , \quad 0 < t)$$

with the conditions:

$$v(\pm 0,t) = v_0(t)$$

$$v(1-0,t) = v(-1 +0,t) = 0$$

$$v(x,0) = 0$$

where $v_0(t)$ is determined by the following

$$\frac{dv_0}{dt} = g + k\, v_x(0,t) \qquad (g,k = \text{constant})$$

and

$$v_0(0) = 0$$

<u>ANSWER.</u>

$$\frac{g}{k}(1-x) - 2\frac{kg}{\nu^2} \sum_{j=1}^{\infty} \frac{e^{-\nu t \alpha_j^2}}{\alpha_j^2(\alpha_j^2 + \frac{k^2}{\nu^2} + \frac{k}{\nu})} \frac{\sin \alpha_j(1-x)}{\sin \alpha_j} \quad (0 < x < 1)$$

where α_j are positive roots of the equation

$$\alpha \tan \alpha \; - \frac{k}{\nu} = 0$$

<u>Transverse Vibrations of an Elastic Beam</u>

Find the solution of the equation

$$u_{xxxx} + u_{tt} = 0$$

with the conditions

$$u(x,0) = u_t(x,0) = 0$$

$$u(0,t) = u_{xx}(0,t) = 0$$

$$u(c,t) = A \sin \omega t, \quad u_{xx}(c,t) = 0 .$$

Show that the resonance frequencies are

$$\omega = \frac{\pi^2 n^2}{c^2}$$

<u>ANSWER</u>.

$$\bar{u}(x,s) = \frac{A}{s^2+\omega^2} \frac{\sin q\, x \sinh qc + \sinh q\, x \sin qc}{2 \sin qc \sinh qc}$$

where

$$q^4 = -s^2 .$$

5. FOURIER TRANSFORMS

Many boundary value problems can be solved by both Laplace and Fourier transforms. It is, however, advantageous to apply the latter for the cases where the initial conditions are prescribed. In fact, the Fourier Transform is a natural generalization of a Fourier series when the interval in consideration becomes infinite.

<u>Theorem</u>. <u>If f(x) satisfies the Dirichlet conditions* (Section 3, Chapter I) in any finite interval and if (x) is absolutely integrable, then</u>

$$\bar{f}(u) = \frac{1}{\sqrt{2\pi}} \int_{-\infty}^{\infty} f(x)\, e^{iux}\, dx \tag{60}$$

<u>is defined for $-\infty < u < \infty$. The inverse formula is</u>

---

*The Dirichlet conditions are entirely unnecessary if the integration is in the sense of Lebesgue and if the convergence in (ii) is considered "almost everywhere." The conditions can be replaced by a weaker one that the function f is of bounded variation in an finite interval as seen from the proof.

$$f(x) = \frac{1}{\sqrt{2\pi}} \int_{-\infty}^{\infty} \bar{f}(u)\, e^{-iux}\, du \tag{61}$$

in the sense that

(i) $f(x) = \dfrac{f(x+0) + f(x-0)}{2}$

(ii) The integral (61) is taken as

$$\lim_{T\to\infty} \frac{1}{\sqrt{2\pi}} \int_{-T}^{T} f(u)\, e^{-iux}\, du$$

First of all, let us establish the following fundamental lemma:

Lemma. (Riemann-Lebesgue). If $f(x)$ is absolutely integrable in $[-\infty, \infty]$, then

$$\lim_{a\to\infty} \int_{-\infty}^{\infty} f(x)\, e^{iax}\, dx = 0 \tag{62}$$

Since $f(x)$ is absolutely integrable, it is clear that for any given ε, there exists a number N large enough such that

$$\left\{\int_{N}^{\infty} + \int_{-\infty}^{-N}\right\} |f(x)| \quad dx < \frac{\varepsilon}{2} \tag{63}$$

Let N be fixed. By a similar argument as in the proof of the lemmas 2 and 3 in Section 7.4, Chapter II, it is readily shown that for any given $\varepsilon > 0$ and a fixed $N > 0$, there exists a polynomial $P(x)$ such that

$$\int_{-N}^{N} |f(x) - P(x)| \quad dx < \frac{\varepsilon}{4}\,. \tag{64}$$

Now both $P(x)$ and $P'(x)$ are bounded on $(-N, N)$ say $|P(x)|$, $|P'(x)| < M$. By integration by parts, one has

$$\int_{-N}^{N} P(x)\, e^{iax}\, dx = \frac{P(x)\, e^{iax}}{a}\Bigg|_{-N}^{N} - \frac{1}{a}\int_{-N}^{N} P'(x)e^{-axi}\, dx$$

This yields the fact that

$$\left|\int_{-N}^{N} P(x)\, e^{iax}\, dx\right| < \frac{2M}{|a|} + \frac{1}{|a|}\, 2NM < \frac{\varepsilon}{4} \tag{65}$$

where

$$|a| > \frac{8(n+1)M}{\varepsilon}$$

combining (63), (64) and (65), the relation (62) follows immediately.

Proof of the Theorem.

Since the infinite integral in (60) converges absolutely and uniformly when $-T \leqq u \leqq T$, the interchange of the order of integration of the expression (ii) is legitimate. Put

$$2\pi F(x,T) = \int_{-T}^{T} e^{-iux}\, du \int_{-\infty}^{\infty} f(\xi)\, e^{iu\xi}\, d\xi$$

$$= \int_{-\infty}^{\infty} f(\xi)\, d\xi \int_{-T}^{T} e^{iu(\xi-x)}\, d\xi$$

$$= 2\int_{-\infty}^{\infty} f(\xi)\, \frac{\sin T\,(\xi-x)}{\xi-x}\, d\xi \tag{66}$$

$$= \int_{-\infty}^{\infty} [f(x+\xi) + f(x-\xi)]\, \frac{\sin T\xi}{\xi}\, d\xi \tag{67}$$

The last equality is obtained by the substitution of $x+\xi$ for ξ and $x-\xi$ for ξ respectively in (66). Since

$$\int_{-\infty}^{\infty} \frac{\sin T\xi}{\xi}\, d\xi = \begin{cases} \pi, & \text{for } T > 0 \\ 0, & \text{for } T = 0 \end{cases}$$

(cf. Ex. 3(a) Section 4, the appendix), it suffices then to show that

$$\lim_{T\to\infty} F(x,T) = \frac{1}{2}\{f(x+0) + f(x-0)\}$$

or

$$\lim_{T\to\infty} \int_0^\infty g(x,\xi)\,\frac{\sin T\xi}{\xi}\,d\xi = 0 \;, \tag{68}$$

where

$$g(x,\xi) = g_1(x,\xi) + g_2(x,\xi)$$

$$g_1(x,\xi) = f(x+\xi) - f(x+0)$$

and

$$g_2(x,\xi) = f(x-\xi) - f(x-0).$$

It is remarked that if $f(\xi)$ satisfies the Dirichlet conditions (or if $f(\xi)$ is of bounded variation in any finite interval), $g_j(x,\xi)$ ($j = 1,2$), considered as a function of ξ, is monotonic (or is a difference of two monotonic functions). Now, $g_1(x,+0) = g_2(x,+0)=0$. By means of the lemma, it remains only to prove that any one of the following integrals.

$$\int_0^\delta g_i(x,\xi)\,\frac{\sin T\xi}{\xi}\,d\xi \qquad (i = 1,2)$$

can be made as small as one pleases provided that is small enough (for any T). Let the integral

$$\int_0^\delta g_1(x,\xi)\,\frac{\sin T\xi}{\xi}\,d\xi$$

be considered. The proof for the other integrals follows by the same arguments. Assume, without loss of generality, that $g_1(x,\xi)$ is monotonically increasing in ξ for $0 \leq \xi \leq \delta$. For a given ε, let be small enough such that $0 \leq g_1(x,\xi) < \varepsilon$. Then, by the second law of the mean for the integral, which is a consequence of Abel's lemma and the definition of the integral, one concludes that

$$K = \int_0^\delta g_1(x,\xi)\,\frac{\sin T\xi}{\xi}\,d\xi = g_1(x,\delta)\int_{\delta_1}^\delta \frac{\sin T\xi}{\xi}\,d\xi$$

where

$$0 < \delta_1 < \delta. \quad \text{Now the integral} \quad \int_{T\delta_1}^{T\delta} \frac{\sin \xi}{\xi}\, d\xi$$

is bounded by a constant M which is independent of T. Therefore

$$|K| < M\varepsilon .$$

This completes the proof.

EXERCISE.

Prove the Theorem in Section 3, Chapter I.

HINT. Show that

(i) The partial sum $S_n(x)$ of the Fourier series is equal to

$$F(x,n) = \int_{-1}^{1} g(x,\xi) \frac{\sin \pi(n + \frac{1}{2})\xi}{2 \sin \frac{1}{2} \pi \xi} d\xi$$

where

$$g(x, \xi) = \frac{1}{2} \{f(x+\xi) + f(x-\xi)\}$$

(ii)

$$\lim_{n\to\infty} [F(x,n) - \int_{-1}^{1} g(x,\xi) \frac{\sin (n + \frac{1}{2})\pi\xi}{\pi\xi} d\xi] = 0$$

Since a continuous function on a closed interval is uniformly continuous the above expression tends uniformly to the limit 0 provided that $f(x)$ is continuous in $[-1, 1]$ and $f(-1) = f(1)$. It should be remarked that the theorem holds for $f(x)$ being of bounded variation instead of satisfying Dirichlet conditions.

5.1 Laplace Inversion Formula as a Special Case

It is immediately shown that if $F(t)\, e^{-s_0 t}$ is integrable in Riemann's sense in $[0, \infty]$, then

$$G = \int_0^t F(t)\, e^{-s_0 t}\, dt$$

is bounded for all values of t, say

$$|G(t)| > M.$$

Then for $Re(s) > s_o$

$$\bar{F}(s) = \int_0^\infty e^{-st} F(t)\, dt = G(t)\, e^{-(s-s_o)t} \Big|_0^\infty$$

$$+ (s-s_o) \int_0^\infty G(t)\, e^{-(s-s_o)t}\, dt$$

$$= (s-s_o) \int_0^\infty G(t)\, e^{-(s-s_o)t}\, dt$$

where the integral converges uniformly for $Re(s-s_o) \geqq \delta > 0$. Hence $\bar{F}(s)$ is analytic in s for $Re(s-s_o) > 0$.

Theorem. <u>If $\bar{F}(s)$ is analytic in the right half plane $Re(s) > s_o$ and if</u>

$$|\bar{F}(s)| < \frac{c}{|s|^k} \quad \text{where } k > 1, \text{ then}$$

$$F(t) = L^{-1}\{\bar{F}(s)\} = \frac{1}{2\pi i} \int_{\gamma-i}^{\gamma+i} e^{st}\, \bar{F}(s)\, ds, \quad (t \geqq 0)$$

<u>where γ is any real constant > s. Moreover F(t) is continuous in t, such that F(0) = 0.</u>

Proof. That (69) is independent of γ is an immediate consequence of Cauchy's theorem. For $t \leqq 0$, consider the contour $\Gamma_1 + \Gamma_2$ as indicated in Fig. 7.

By Cauchy's theorem, $\oint_\Gamma e^{st} F(s)\, ds = 0$. Since on Γ_1

$$|\oint_\Gamma e^{st} F(s)\, ds| \leqq \int_{-\frac{\pi}{2}}^{\frac{\pi}{2}} e^{tR\cos\theta} \frac{R}{R^k}\, d\theta \leqq \frac{\pi}{R^{k-1}}$$

Thus as $R \to \infty$, F(t) = 0. (70)

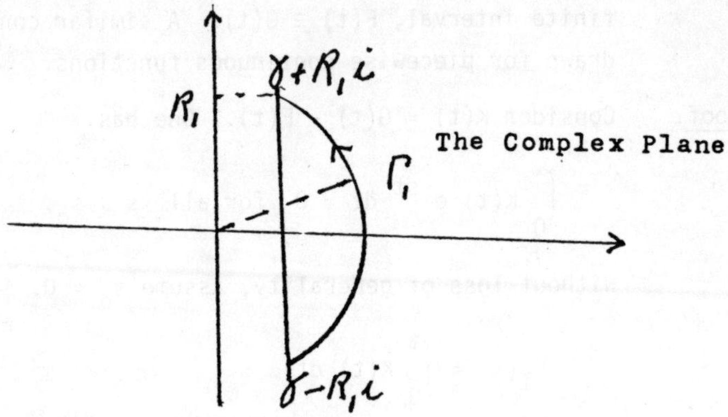

Figure 9

On the other hand, (69) converges uniformly and absolutely for $c \geqq t \geqq 0$. $F(t)$ is therefore continuous. Now, let $s = \gamma+vi$. One has

$$\sqrt{2\pi}\, F(t)\, e^{-\gamma t} = \frac{1}{\sqrt{2\pi}} \int_{-\infty}^{\infty} e^{ivt}\, \bar{F}(\gamma+iv)\, dv \tag{71}$$

By the Fourier inversion theorem in Section 5, one concludes from (70) and (71)

$$F(s) = \int_0^{\infty} F(t)\, e^{-st}\, dt \tag{72}$$

This completes the proof.

5.2. Uniqueness of the Inverse Laplace Transform

Theorem. If $F(t)$ and $G(t)$ have the same Laplace transform for $s > s_0$, then

$$\int_0^t F(t)\, dt \equiv \int_0^t G(t)\, dt \tag{73}$$

for all values of $t \geqq 0$.

If, in addition, both $F(t)$ and $G(t)$ are continuous in any finite interval, $F(t) \equiv G(t)$. A similar conclusion can be drawn for piecewise continuous functions.

Proof. Consider $K(t) = G(t) - F(t)$. One has

$$\int_0^\infty K(t)\, e^{-st}\, dt = 0 \quad \text{for all} \quad s \geqq s_o.$$

Without loss of generality, assume $s_o = 0$. Let

$$K_1(t) = \int_0^t K(t)\, dt$$

Consequently, $|K_1(t)| < M$. It follows by integration in parts that

$$\int^\infty K_1(t)\, e^{-st}\, dt = 0 \qquad (s > 0) \tag{74}$$

Put $x = e^{-t}$ and $K_1\,(-\log x) = K_2(x)$ where

$$K_2(0) = \lim_{t\to\infty} K_1(t).$$

Thus $K_2(x)$ is continuous in $[0,1]$. Now the equation (74) becomes

$$\int_0^1 K_2(x)\, x^{s-1}\, dx = 0 \tag{75}$$

This relation holds, in particular for $s = 1,2, \ldots$.
By exactly the same reasoning, it is clear that

$$\int_0^1 K_3(x)\, x^n\, dx = 0 \qquad (n = 0,1,2 \ldots)$$

where

$$K_3(x) = \int_0^x K_2(x)\, dx.$$

Now by the theorem in Section 5.1, Chapter II,

$K_3(x) \equiv 0$. So is $K_2 \equiv 0$.

Consequently,

$$K_1(t) \equiv 0 .$$

This completes the proof.

5.2.1 Elementary Rules. Assume that all the functions which appear below are absolutely integrable. Let $F\{f(x)\} \equiv F(f)$, be the Fourier transform of $f(x)$. Let

$$F_c\{f(x)\} = F_c\{f\} \quad \text{and} \quad F_s\{f(x)\} = F_s\{f\}$$

be respectively the Fourier cosine and the Fourier sine transforms of $f(x)$. The following rules are immediately derived by the same reasoning as for those of Laplace Transforms.

$$\text{(i)} \quad F\{f'(x)\} = -i\sigma F\{f\} \tag{76}$$

$$\text{(ii)} \quad F\{h(x)\} = F\{f(x)\} \cdot F\{g(x)\} \tag{77}$$

where

$$h(x) = \frac{1}{\sqrt{2\pi}} \int_{\infty}^{\infty} f(x-\xi)g(\xi)\, d \;\; = \frac{1}{\sqrt{2\pi}} \int_{-\infty}^{\infty} f(\xi)g(x-\xi)\, d\xi \tag{78}$$

$$\text{(iii)} \quad F_s\{F_c\{f\}\, F_s\{g\}\} = \frac{1}{\sqrt{2\pi}} \int_0^{\infty} g(\xi)\, [f(|x-\xi|) - f(x+\xi)]d\xi \tag{79}$$

$$\text{(iv)} \quad F_c\{F_c\{f\} G_c\{g\}\} - \frac{1}{\sqrt{2\pi}} \int_0 g(\xi)[f(|x-\xi|) + f(x+\xi)]\, d\xi \tag{80}$$

The proofs of these formulas are left for the student as exercises. The function $e^{-x^2/2}$ can be verified readily, by complex integration, to be an invariant under the Fourier transformation:

$$F_c\{e^{-\frac{x^2}{2}}\} = F\{e^{-\frac{x^2}{2}}\} = e^{-\frac{\sigma^2}{2}} \tag{81}$$

6. THE SOLUTION OF BOUNDARY VALUE PROBLEMS BY FOURIER TRANSFORMS

Consider the heat conduction in a semi-infinite bar. The equation is

$$u_t = u_{xx} \quad (t > 0 \qquad 0 < x < \infty)$$

with the boundary conditions:

$$u(x,+0) = f(x) \qquad (0 < x < \infty)$$

$$u(+0,t) = 0 \qquad (t > 0)$$

$$\lim_{x\to\infty} u(x,t) = 0 \qquad (t > 0)$$

For such a problem one, of course, can define

$$u(-x,t) = -u(x,t)$$

and then apply the Fourier sine transform. Put

$$\bar{U}(\sigma,t) = F_s \{u(x,t)\}$$

It is readily verified that

$$F_s \{u_{xx}(x,t)\} = - F_c \{u_x(x,t)\}$$

$$= -\sigma u(x,t) \cos \sigma x \Big|_{x=0}^{x=\infty} -\sigma^2 F_s \{u(x,t)\}$$

$$= -\sigma^2 \bar{U}(\sigma,t)$$

Thus one has formally

$$\bar{U}_t = -\sigma^2 \bar{U} , \quad \bar{U}(\sigma,+0) = F_s\{f\}$$

This gives

$$\bar{U}(\sigma,t) = F_s \{f\} e^{-\sigma^2 t}$$

Now by virtue of (81)

$$e^{-\sigma^2 t} = F_c\{ \frac{1}{2t} e^{-\frac{x^2}{4t}}\} \tag{82}$$

Therefore by (79), one finds the solution

$$u(x,t) = \frac{1}{2\sqrt{\pi t}} \int_0^\infty f(\xi) \left[e^{-\frac{(x-\xi)^2}{4t}} - e^{-\frac{(x+\xi)^2}{4t}} \right] d\xi \tag{83}$$

A physical interpretation which is rather important is that the temperature after time t at x due to a source initially at ξ is

$$\frac{1}{2\sqrt{\pi t}} e^{-\frac{(x-\xi)^2}{4t}}$$

The solution is then obtained by "the method of images" with sources and sinks of equal strength $f(\xi)$, placed symmetrically with respect to the end $x = 0$.

Ex 1. Application to Surface Waves

Consider a two-dimensional incompressible fluid. Let the y-axis be vertical and the x-axis horizontal. Let $\varphi(x,y,t)$ be the velocity potential, which is defined up to an arbitrary function of t, such that

$$u = \text{the x-component of the velocity} = \varphi_x$$

$$v = \text{the y-component of the velocity} = \varphi_y$$

We have the Bernoulli's equation

$$\frac{p}{\rho} = \frac{\partial \varphi}{\partial t} - gy \quad \text{(when the term involving } \frac{u^2+v^2}{2} \text{ is neglected).} \tag{84}$$

where ρ = constant and g is the gravitation constant. Let $y = 0$ be the level of the liquid at rest. Let $y = \eta(x,t)$ be the elevation of the free surface. From (84),

$$\eta = \frac{1}{g} \varphi_t(x,\eta,t) - \frac{1}{g\rho} p_o(x,t) \tag{85}$$

where $p_o(x,t)$ designates the pressure at surface. Or, the approximation of (85) is

$$\eta(x,t) = \frac{1}{g} \varphi_t(x,o,t) - \frac{1}{g\rho} p_o(x,t) \tag{85'}$$

Since

$$v(x,o,t) = \frac{\partial\eta}{\partial t} = \frac{1}{g}\,\varphi_y(x,o,t) \tag{86}$$

Hence we have the boundary condition when the surface is free ($p_o = 0$)

$$\varphi_{tt}(x,o,t) + g\,\varphi_y(x,o,t) = 0 \tag{87}$$

On the other hand, if an impulse pressure is applied to the free surface, for a short time interval we have

$$\eta = 0\ ,\quad \varphi_t(x,o,t) = \frac{1}{\rho}\,p_o(x,t)$$

or

$$\varphi(x,o,o) = \frac{P_o(x)}{\rho} = \lim_{t\to o}\int_o^t p_o(x,t)\,dt/\rho \tag{88}$$

Thus, we have the boundary value problem:

$$\varphi_{xx} + \varphi_{yy} = 0$$

besides (87) and (88), φ is bounded as $y \to -\infty$. Find the solution of such a problem when $P_o(x) = P\delta(x)$ where P is a constant.

ANSWER.

$$\varphi(x,y,t) = \frac{P}{\rho 2\pi}\,\Big[\int_o^\infty e^{\xi y}\cos(\xi x - \sqrt{g\xi}\,t)\,d\xi$$

$$+ \int_o^\infty e^{\xi y}\cos(\xi x + \sqrt{g\xi}\,t)\,d\xi$$

Ex. 2.

Show that the elevation of the free surface in Ex. 1 is

$$\eta(x,t) = \frac{P}{\rho(2\pi g x^3)^{\frac{1}{2}}}\,\{C(\mu)\,[\cos\frac{\pi\mu^2}{2} - \pi\mu^2\sin\frac{\pi\mu^2}{2}]$$

$$+ S(\mu)\,[\sin\frac{\pi\mu^2}{2} + \pi\mu^2\cos\frac{\pi\mu^2}{2}] + 1\}$$

where $C(\mu)$ and $S(\mu)$ are Fresnel's integrals defined by

$$C(z) = \int_0^z \cos \frac{\pi x^2}{2}\, dx \ , \quad S(z) = \int_0^z \sin \frac{\pi x^2}{x}\, dx,$$

where

$$(\mu = gt^2/2\pi x)$$

The behavior of functions are indicated in the following figures:

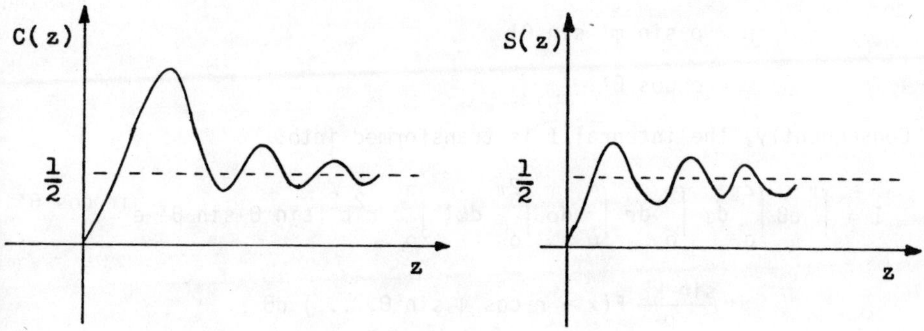

Fig. 10

6.1 Remarks on Multiple Fourier Transforms

Repeated Fourier transformations with respect to space coordinates are often applied to the problems when the medium occupies the whole space. In the isotropic case, the evaluation of Fourier transforms is facilitated by the use of spherical coordinates. As an illustration, consider the following integral:

$$I = \iiiiiint_{-\infty}^{\infty} F(\alpha,\beta,\gamma)\, \frac{\sin k\rho}{\rho}\, e^{i\,[\lambda(\alpha-x)+\mu(\beta-y)+\nu(\gamma-z)]} \, d\alpha\, d\beta\, d\gamma\, d\lambda\, d\mu\, d\nu \tag{89}$$

where (x,y,z) and k are fixed parameters, and

$$\rho^2 = \lambda^2 + \mu^2 + \nu^2 \ .$$

Let

$$\alpha - x = r \cos \varphi \sin \theta$$

$$\beta - y = r \sin \varphi \sin \theta$$

$$\gamma - z = r \cos \theta$$

so that (r,φ,θ) are spherical coordinates of (α,β,γ) with respect to an origin at (x,y,z). Again, for the point (α,β,γ), we take, for polar axis, from which θ' is measured, the line joining (x,y,z) to α, β, γ

$$\lambda = \rho \cos \varphi' \sin \theta'$$

$$\mu = \rho \sin \varphi' \sin \theta'$$

$$\nu = \rho \cos \theta'.$$

Consequently, the integral I is transformed into:

$$I = \int^{\pi} d\theta \int_{0}^{2\pi} d\varphi \int_{0}^{\infty} dr \int_{0}^{\infty} d\rho \int_{0}^{2\pi} d\varphi' \int_{0}^{\pi} r^2\rho^2 \sin\theta \sin\theta' e^{ir\rho\cos\theta'}$$

$$\cdot \frac{\sin k\rho}{\rho} F(x + r\cos\psi\sin\theta, \ldots)\, d\theta$$

$$= 4\pi \int_{0}^{2\pi} d\varphi \int_{0}^{\pi} \sin\theta\, d\theta \int_{0}^{\infty} d\rho \sin k\rho \int_{0}^{\infty} r \sin r\rho F(x$$

$$+ r\cos\varphi\sin\theta, \ldots)dr$$

$$= 8\pi^2 \int_{0}^{2\pi} d\varphi \int_{0}^{\pi} \sin\theta\, d\theta\, [kF(x+k\cos\varphi\sin\theta, \ldots)] \qquad (90)$$

where the last equality follows the formula for Fourier sine transforms. The integral (90) can be considered as one over the surface of unit sphere. The following Cauchy's solution for the initial value problem of the sound wave in space is obtained by means of the remarks just made and is left to the reader as an exercise.

Ex.

Show, by means of multiple Fourier transforms, that the wave equation

$$\Delta U - \frac{1}{a^2} U_{tt} = 0 \quad (-\infty < x,y,z < \infty \qquad t > 0)$$

with the initial conditions that

$$U(x,y,z,o) = f(x,y,z)$$

$$u_t(x,y,z,o) = g(x,y,z)$$

is

$$U = \frac{t}{4\pi}\int_o^{\pi}\int_o^{2\pi} g(x+at\cos\varphi\sin\theta, \ldots)\sin\theta\, d\theta\, d\varphi$$

$$+ \frac{\partial}{\partial t}\left[\frac{t}{4\pi}\int_o^{\pi}\int_o^{2\pi} f(x+at\cos\varphi\sin\theta, \ldots)\sin\theta\, d\theta d\varphi\right. \quad (91)$$

The interpretation of (91) is clear that, if a disturbance is initially generated in a bounded domain D in the space, it takes some time to reach a point P (not in D) and has the influence at the point P for some time period. After that, the disturbance disappears as far as this point is concerned.

CHAPTER IV.

OPERATIONAL METHOD II. OTHER INTEGRAL TRANSFORMS

The present treatment of the subject is purely formal. Particular interests will be centered at general integral transforms with Fourier kernels. Up to now, we have employed Fourier and Laplace transforms as operational methods to solve the boundary value problems of partial differential equations. In the several concrete examples given later, we shall employ other integral transforms to reduce some boundary value problems into a pair of integral equations, the formal solution of which is obtained by Titchmarsh.*

In applying such a procedure to a boundary value problem, neither do we employ, at any time, the conditions of validity for transforms nor shall we care about it at the moment. As a matter of fact, if such an operational method should yield an explicit solution of the problem, to establish it as a rigorous solution is often a matter of verification.

These realistic excuses, by no means, constitute an argument against the very spirit of Modern Mathematics--the Mathematical rigor. On the contrary, the conscientious reader is reminded of the intrinsic need and importance of rigorous treatment.

1. INTEGRAL TRANSFORMS AND THEIR INVERSE

Let the general integral transform of an arbitrary function f belonging to a certain class of functions A* be formally defined as:

$$f^*(x) = \int_a^b K(x,y)\, f(y)\, dy \tag{1}$$

where K(x,y) is called the kernel of the transform; a and b are constants. Observe that the integral is performed with respect to the second variable.

---

*E.C. Titchmarsh, Theory of Fourier Integral, pp. 337-339. Also I. Busbridge, Proc. London Math Soc. 44, p. 115, (1938).

All these f* form another class of function A*. If for any f* in A*, there is a kernel H(y,x) such that:

$$f(y) = \int_c^d H(y,x)\, f^*(x)\, dx \tag{2}$$

then H(y,x) is called the kernel for the inverse transforms.

As for illustrations, we mention the following examples:

(1) Laplace Transforms

$$K(x,y) = e^{-xy}; \quad a = 0, \quad b = +\infty$$

$$H(y,x) = \frac{1}{2\pi i} e^{xy}; \quad c = \gamma - i\infty, \quad d = \gamma + i\infty$$

(2) Fourier Transforms

$$K(x,y) = e^{-ixy}; \quad a = -\infty, \quad b = +\infty$$

$$H(y,x) = \frac{1}{2\pi} e^{-ixy}; \quad c = -\infty, \quad d = +\infty.$$

(3) Mellin Transforms

$$K(x,y) = y^{x-1}; \quad a = 0, \quad b = +\infty$$

$$H(y,x) = \frac{1}{2\pi i} y^{-x}; \quad c = \gamma - i\infty, \quad d = \gamma + i\infty$$

The inversion formula for Mellin transforms can be derived formally from Fourier transforms as follows:

Let: $x = \gamma + iv$ where γ is kept constant and $y = e^t$. We have:

$$f^*(\gamma+iv) = \int_0^\infty f(y)\, y^{x-1}\, dy$$

$$= \int_{-\infty}^\infty f(e^t)\, e^{\gamma t} \cdot e^{ivt}\, dt$$

by Fourier inversion formula:

$$f(e^t)\, e^{\gamma t} = \frac{1}{2\pi} \int_{-\infty}^\infty f^*(\gamma+vi)\, e^{-ivt}\, dv$$

Hence, dividing by $e^{\gamma t}$ and transforming back to the original variables, we have:

$$f(y) = \frac{1}{2\pi i}\int_{\gamma-i\infty}^{\gamma+i\infty} f^*(x)\, y^{-x}\, dx$$

The conditions for validity of this inversion formula can be derived readily from Fourier transforms.

(3.1) Convolution Formulas for Mellin Transforms

The following convolution formulas are extremely useful. The formal derivation is given here. They are valid under various conditions and in various senses, although only one set of conveniently stated conditions are mentioned.

If $f(y)y^k$ and $g(y)y^{\sigma-k}$ are both square integrable, then for $x = \sigma+it$

$$\int_0^\infty f(y)\, g(y) y^{x-1} dy = \frac{1}{2\pi i}\int_{k-i\infty}^{k+i\infty} f^*(\xi)\, g^*(x-\xi)\, d\xi \qquad (4)$$

The "formal" verification is following:

$$\int_0^\infty g(y)y^{x-1}dy \, \frac{1}{2\pi i}\int_{k-i\infty}^{k+i\infty} f^*(\xi)\, y^{-\xi} d\xi$$

$$= \frac{1}{2\pi i}\int_{k-i\infty}^{k+i\infty} f^*(\xi)\, d\xi \int_0^\infty g(y)y^{(x-\xi)-1}\, dy$$

Hence (4) follows immediately.
Similarly, we have:

$$\frac{1}{2\pi i}\int_{k-i\infty}^{k+i\infty} f^*(\xi)\, g^*(\xi) y^{-\xi} d\xi = \int_0^\infty f\left(\frac{y}{u}\right) g(u)\, \frac{du}{u}$$

provided both f^* and g^* are square integrable on the line $\mathrm{Re}(\xi) = k$.

Its formal verification is left to the reader.

EXERCISES

(1) Find the Mellin Transforms for sin y and cos y.

ANSWER.

$$\sin\frac{x\pi}{2}\,\Gamma(x), \quad \cos\frac{x\pi}{2}\,\Gamma(x)$$

HINT. The Γ-function is defined as follows:

$$\Gamma(x) = \int_0^{\infty} e^{-t}\, t^{x-1}\, dt$$

Show that

(1) $$\int_0^{\infty} e^{-py}\, y^{x-1} dy = \frac{\Gamma(x)}{p^x}$$

(2) Set $p = i = e^{i\pi/2}$. When x is real, find the real and imaginary parts of the integral.

(2) Let the Mellin transform of $\sqrt{\frac{2}{\pi}}\sin y$ or $\sqrt{\frac{2}{\pi}}\cos y$ be $K(x)$, show that:

$$K(x)\, K(x-1) \equiv 1$$

HINT.

(1) Prove that:

$$I = \Gamma(x)\,\Gamma(1-x) = 2\int_0^{\pi/2} \tan^{2x-1}\theta\, d\theta = 2\int_0^{\infty} \frac{\xi^{2x-1}}{1+\xi^2}\, d\xi$$

(2) Consider the complex integration

$$\frac{1}{2\pi i}\oint_C \frac{z^{2x-1}}{1+z^2}\, dz$$

where C is a closed contour consisting of the segment from -R to R along the real axis and the half circle joining R to -R in the upper-half plane.

Define:

$$z^k = e^{k(\log\rho + i\theta)}$$

where:

$$z = \rho e^{i\theta} \quad\text{and}\quad -\frac{\pi}{2} < \theta < \frac{3\pi}{2}$$

Thus, show that the integral I in (1) is:

$$I = \frac{\pi}{\sin(x\pi)}$$

2. FOURIER KERNELS

We are concerned with the special case that a = c = 0, and b = d = ∞ and:

$$K(x,y) = \varphi(xy)$$

$$H(y,x) = \psi(xy)$$

Such kernels are called <u>Fourier kernels</u>. Here φ and ψ are both functions of one variable.

<u>Theorem:</u> Let the Mellin transforms of φ and ψ be respectively Φ and Ψ. Finally, φ and ψ are Fourier kernels if and only if:

$$\Phi(s)\,\Psi(1-s) \equiv 1 \ .$$

<u>Proof</u>. The necessary condition. Suppose:

$$f^*(x) = \int_0^\infty \varphi(xy)\, f(y)\, dy \tag{6}$$

and:

$$f(y) = \int_0^\infty \psi(xy)\, f^*(x)\, dx \tag{7}$$

Let the Mellin transforms of the function be designated by the capital letters. Taking the Mellin transforms of (6) and (7), we have:

$$F^*(s) = \Phi(s)\, F(1-s) \tag{6'}$$

and:

$$F(s) = \Psi(s)\, F^*(1-s) \tag{7'}$$

Changing s into 1-2 in (7'), we have:

$$F(1-s) = \Psi(s)\, F^*(s) \tag{7"}$$

Then, from (6') and (7'), we have (5)

In order to show formally that the condition is also sufficient, we shall first establish, formally, the following:

Lemma. Let $\varphi_1(t) = \int_0^t \varphi(\xi)\, d\xi$. A necessary and sufficient condition that φ and ψ form Fouriers kernels is that:

$$\int_0^\infty \frac{\psi(xu)\,\varphi_1(\xi u)}{u}\, du = \begin{cases} 1, & (0 < x < \xi) \\ 0, & (x > \xi) \end{cases} \tag{8}$$

That (8) is a necessary condition is readily seen as follows:

Take, for any fixed ξ,

$$f(x) = \begin{cases} 1, & (0 < x < \xi) \\ 0, & (x < \xi) \end{cases}$$

The general inversion formula yields:

$$f(x) = \int_0^\infty \psi(xu)du \int_0^\infty \varphi(uy)\, f(y)dy$$

$$= \int_0^\infty \psi(xu)\, du \int_0^\xi \varphi(uy)dy$$

$$= \int_0^\infty \frac{\psi(xu)\varphi_1(\xi u)}{u}\, du$$

which is (8).

On the other hand, if (8) holds, we have for an arbitrary $f^*\alpha$,

$$\int_0^\xi f^*(x)dx = \int_0^\infty f^*(x)dx \int_0^\infty \frac{\psi(xu)\varphi_1(\xi u)}{u}\, du$$

$$= \int_0^\infty \varphi_1(\xi u)du \int_0^\infty \frac{f^*(x)\psi(xu)}{u}\, dx$$

A formal differentiation gives the inversion formula for the general transforms.

This completes the formal proof of the lemma.

Returning to the proof of the sufficiency of (5), we have

$$\varphi(y) = \frac{1}{2\pi i}\int_{k-i\infty}^{k+i\infty} \Phi(s)y^{-s}\, ds$$

Hence:

$$\varphi_1(y) = \frac{1}{2\pi i}\int_{k-i\infty}^{k+i\infty} \frac{\Phi(s)y^{-s+1}}{1-s}\, ds \tag{10}$$

Consequently, by means of (10) and interchanging the order of integration, it follows:

$$\int_0^\infty \frac{\psi(xu)\varphi_1(\xi u)}{u}\, du$$

$$= \frac{1}{2\pi i}\int_{k-i\infty}^{k+i\infty} \frac{\Phi(s)}{1-s}\,\xi^{1-s}\, ds \int_0^\infty \psi(xu)\, u^{-s}\, du$$

$$= \frac{1}{2\pi i}\int_{k-i\infty}^{k+i\infty} \frac{\Phi(s)\Psi(1-s)}{1-s}\left(\frac{\xi}{x}\right)^{1-s} ds$$

$$= \frac{1}{2\pi i}\int_{k-i\infty}^{k+i\infty} \frac{a^{1-s}}{1-s}\, ds$$

Where $a = \frac{\xi}{x}$, and the last equality is form (5). By the residue theory, it follows readily that the value of the integral is:

$$\begin{cases} 1, & \text{for } a > 1 \quad \text{or} \quad x < \xi \\ 0, & \text{for } a < 1 \quad \text{or} \quad x > \xi . \end{cases}$$

This completes the formal proof of the theorem.

2.1 Self-Associated Fourier Kernels

When $\varphi = \psi$, we shall call the Fourier kernel as self-associated. From Ex. 2 of Section 1, it is clear that both sine and cosine functions are self-associated Fourier kernels. Further examples can be easily constructed from the Mellin transforms of the kernels.

2.1.1 Hankel Transforms. An example, let us consider the Fourier kernel φ, whose Mellin transform is:

$$\Phi(s) = 2^{s-\frac{1}{2}} \frac{\Gamma(\frac{1}{2}\nu + \frac{1}{2}s + \frac{1}{4})}{(\frac{1}{2}\nu - \frac{1}{2}s + \frac{3}{4})} \tag{11}$$

where ν is a constant.

From the above theorem in Section 2, the Mellin transform of its associated kernel is:

$$\Psi(s) = \frac{1}{\Phi(1-s)}$$

$$= \Phi(s)$$

Hence it is a self-associated Fourier kernel.

We shall establish the result that the inverse Mellin transform of (11) is:

$$\sqrt{x}\, J_\nu(x)\ .$$

Thus, formerly we have the following:

Theorem. If:

$$f^*(x) = \int_0^\infty \sqrt{xy}\, J_\nu(xy)\, f(y)\, dy \tag{12}$$

then:

$$f(y) = \int_0^\infty \sqrt{xy}\, J_\nu(xy)\, f^*(x)\, dx \tag{13}$$

These formulas can be established rigorously under various conditions and in various senses for $\nu > -\frac{1}{2}$, for which we are not going to indulge there. If we set:

$$\frac{f(y)}{\sqrt{y}} = g(y)$$

and:

$$\frac{f^*(x)}{\sqrt{x}} = g^*(x)$$

we have the alternative formula:

$$g(y) = \int_0^\infty J_\nu(yu)\ udu \int_0^\infty J_\nu(xu)\ xg(x)\ dx \qquad (14)$$

The form of (14) will be employed for the future discussions.

2.1.2 Some basic facts concerning Γ-function. Before we are going to find the inverse Mellin transform of (11), a few basic facts concerning the Γ-function as a complex function should be mentioned. In the Ex. of Section 1, we have defined Γ(x) when Re(x) > 0.

This function considered as an analytic function of a complex variable is now continued analytically into the whole plane.

The infinite product:

$$g(z) = \prod_{n=1}^{\infty} (1 + \frac{z}{n})\ e^{-\frac{z}{n}}$$

converges uniformly to the nonvanishing function for z in any bounded, closed domain excluding negative integers. Consequently, the function, defined by Weierstrass:

$$f(z) = \frac{1}{zg(z)}\ e^{-\gamma z}\ , \quad \text{where} \quad \gamma = \lim_{m\to\infty} [\sum_{n=1}^{m} \frac{1}{n} - \log m]$$

is a meromorphic function of z.

We are going to show that f(z) = Γ(z) for Re(z) > 0. First of all, let us derive Euler's formula:

$$\frac{1}{f(z)} = z \lim_{m\to\infty} [e^{z(1+\frac{1}{2}+...+1/m-\log m)} \prod_{n=1}^{m} (1 + \frac{z}{n})\ e^{-\frac{z}{n}}]$$

$$= z \lim_{m\to\infty} [m^{-z} \prod_{n=1}^{m} (1 + \frac{z}{n})]$$

Hence,

$$f(z) = \lim_{n\to\infty} \frac{1,2...(n-1)}{z(z+1)...(z+n-1)}\ n^z$$

It follows immediately from Euler's formula that:

$$f(z+1) = zf(z) \quad \text{for all} \quad z \neq \text{nonpositive integer} \qquad (15)$$

Now, denote:

$$\pi(z,n) = \int_0^n (1 - \frac{t}{n})^n t^{z-1}\, dt$$

Integrating by parts, it follows that:

$$\pi(z,n) = \frac{n!}{z(z+1)\dots(z+n-1)} \int_0^n \frac{t^{z+n-1}dt}{n^n}$$

By Euler's formula, we have:

$$\lim_{n\to\infty} \pi(z,n) = f(z).$$

On the other hand, we are going to prove that $f(z) = \Gamma(z)$ when $z > 0$. Consider:

$$\Gamma(z) - f(z) = \lim_{n\to\infty} [\int_0^n \{e^{-t} - (1 - \frac{t}{n})^n\} t^{z-1} dt + \int_n^\infty e^{-t}\, t^{z-1}\, dt]$$

It is clear that when $z > 0$, the limit of the second integral is zero. The proof that the limit of the first integral is zero is based on the following:

<u>Lemma</u>.

$$0 \leqq e^{-t} - (1 - \frac{t}{n})^n \leqq n^{-1}\, t^2\, e^{-t} \qquad (16)$$

<u>Proof:</u> For any real x between 0 and 1, we have:

$$1 + x \leqq e^x \leqq \frac{1}{1-x} \text{ and } (1-x)^n \geqq 1 - nx.$$

The second inequality is proved by the fact that the x-derivative of $(1-x)^n - 1 + nx$ is non-negative. Hence, it is monotonically increasing in x. Hence, setting $x = \frac{t}{n}$, the first inequality yields:

$$(1 + \frac{t}{n})^{-n} \geqq e^{-t} \geqq (1 - \frac{t}{n})^n$$

Then:

$$0 \leqq e^{-t} - (1 - \frac{t}{n})^n \leqq e^{-t} [1 - (1 - \frac{t^2}{n^2})^n] \leqq e^{-t} \frac{t^2}{n}$$

This completes the proof of the lemma, and consequently, we establish the fact that when $z > 0$, $f(z) = \Gamma(z)$. Thus, the function $\Gamma(z)$ is now extended to the whole complex plane.

Corollary. $\Gamma(z)$ is a nonvanishing meromorphic function of z. When $z = 0$ or a negative integer, it has a simple pole. Finally, if $m = 0$ or a negative integer:

$$\lim_{z\to m} \Gamma(z)\,(z-m) = \frac{(-1)^m}{|m|!}$$

The last statement is based on the fact that:

$$(1)\quad \lim_{z\to 0} z\Gamma(z) = 1 \tag{17}$$

$$(2)\quad \Gamma(m+z_0) = \frac{1}{m+z_0}\cdot\Gamma(1+m+z_0) \tag{18}$$

$$= \frac{1}{(m+z_0)(1+m+z_0)\ldots(-1+z_0)}\,\Gamma(z_0)\ .$$

2.2 Inverse Mellin Transform of (11).

Let:

$$f(x) = \frac{1}{2\pi i}\int_{k-i\infty}^{k+i\infty} \frac{\Gamma(\frac{\nu}{2}+\frac{s}{2}+\frac{1}{4})}{\Gamma(\frac{\nu}{n}-\frac{s}{2}+\frac{3}{4})}\, 2^{s-\frac{1}{2}}\, x^{-s}\, ds \tag{11}$$

This can be evaluated by means of the residue theory. The rigorous justification is based on the asymptotic behavior of Γ-function. We shall not indulge here. The integral is, by the residue theory, the sum of the residues of the integrand at the poles of $\Gamma(\frac{\nu}{2}+\frac{s}{2}+\frac{1}{4})$; namely, $s = -2m - \nu = \frac{1}{2}$ where $m = 0,1,2, \ldots$. Thus,

$$f(x) = \sum_{m=0}^{\infty} R_m$$

where:

$$R_m = \frac{2^{-(2m+\nu+1)}\, x^{(2m+\nu+\frac{1}{2})}}{\Gamma(m+\nu+1)} \lim_{s\to(-2m-\nu-\frac{1}{2})} (s+2m+\nu+\tfrac{1}{2})\,\Gamma(\tfrac{\nu}{2}+\tfrac{s}{2}+\tfrac{1}{4})$$

$$= \frac{(-1)^m\, x^{\frac{1}{2}}}{m!\,\Gamma(m+\nu+1)}\left(\frac{x}{2}\right)^{2m+\nu}$$

Thus, by the expansion of $J_\nu(x)$ as obtained in Chapter II, we have:

$$f(x) = x^{\frac{1}{2}} J_\nu(x) .$$

Consequently, the inversion formula (14) of Hankel transforms is formally established.

2.3 Hankel's Transforms of the Derivative

The following formula is essential in reducing some problems with rotational symmetry into Titchmarsh's dual integral equations:

$$\int_0^\infty [f''(r) + \frac{1}{r} f'(r) - \frac{\nu^2}{r^2} f(r)] J_\nu(r\xi) \, dr$$

$$= -\xi^2 f^*(\xi)$$

$$= -\xi^2 \int_0^\infty f(t) \, t \, J_\nu(t\xi) \, dt . \qquad (19)$$

By means of integration in parts we have:

$$\int_0^\infty [rf''(r) + f'(r)] J_\nu(\xi r) \, dr = -\xi \int_0^\infty rf'(r) \, J'_\nu(\xi,r) \quad dr$$

$$= \xi \int_0^\infty f(r) \frac{d}{dr} [r \, J'_\nu(\xi r)] \, dr$$

$$= -\xi^2 \int_0^\infty f(r) \, rJ_\nu(\xi r) \, [1 - \frac{\nu^2}{\xi^2 r^2}] \, dr \qquad (19')$$

The last equality follows from the Bessel's equation that if $\xi r = z$ and $J_\nu(\xi,r) = y$, then:

$$\frac{d}{dz} (z \frac{dy}{dz}) + (z - \frac{\nu^2}{z}) \, y = 0$$

Hence, (19) follows immediately from (19').

2.4 Example

Steady, incompressible flow through a circular aperture. With the center of the aperture as origin and the direction perpendicular to the plane of aperture as z-axis, the velocity potential $\varphi(r,z)$

satisfies the Laplace equation in cylindrical coordinates (r,z):

$$\varphi_{rr} + \frac{1}{r}\varphi_r + \varphi_{zz} = 0$$

Let the radius of the aperture be 1. Then, when $r > 1$,

$$\varphi_z(r,0) = 0 .$$

Consider the problem that:

$$\varphi(r,0) = g(r),$$

where $g(r)$ is a given function of r. Also, the condition at infinity requires

$$\lim_{z\to+\infty} \varphi(r,z) = 0 .$$

Designate $\varphi^*(s,z)$ the Hankel transform of φ with respect to r. Hence,

$$-s^2\varphi^* + \varphi^*_{zz} = 0 .$$

By virtue of (20), we have:

$$\varphi^* = A(s)\, e^{-sz}$$

Thus:

$$\varphi(r,z) = \int_0^\infty A(s)\, e^{-sz}\, sJ_\nu(sr)\, ds .$$

The conditions (18) and (19) yield the pair of integral equations of the forms:

$$\int_0^\infty x^\alpha f(x)\, J_\nu(xr)\, dx = g(r) \qquad (0 \leq r \leq 1)$$

$$\int_0^\infty f(x)\, J_\nu(xr)\, dx = 0 \qquad (r > 1)$$

where $\alpha = -1$ and $f(x) = x^2 A(x)$.

The formal solution of the dual integral equations is due to Tritchmarsh which is derived in the next section.

3. TITCHMARSH'S DUAL INTEGRAL EQUATIONS

The Mellin transform with respect to y of $y^{\alpha}J_{\nu}(ry)$ by virtue of (11) is:

$$\frac{2^{\alpha+s-1}\,\Gamma(\frac{\alpha+\nu+s}{2})}{r^{\alpha+s}\,\Gamma(\frac{2-\alpha-s+\nu}{2})}$$

Applying the convolution formula (4) to the dual integral equations (18') and (19), we have:

$$\frac{1}{2\pi i}\int_{k-i\infty}^{k+i\infty} \frac{\Gamma(\frac{1+\nu+s}{2})}{\Gamma(\frac{1+\nu+s-\alpha}{2})}\,\chi(s)\,r^{s-1-\alpha}\,ds = g(r) \qquad (0 < r < 1) \tag{20}$$

$$\frac{1}{2\pi i}\int_{k-i\infty}^{k+i\infty} \frac{\Gamma(\frac{1+\nu+s}{2})}{\Gamma(\frac{1+\nu+\alpha-s}{2})}\,\chi(x)\,r^{s-1}\,ds = 0 \qquad (r > 1) \tag{21}$$

where:

$$\chi(s) = 2^{\alpha-s}\,\frac{\Gamma(\frac{1+\nu+\alpha-s}{2})}{\Gamma(\frac{1+\nu+s}{2})}\,f^{*}(s)\ ,$$

$f^{*}(s)$ is Mellin transform of f,

and:

k is a constant.

Multiply (20) by r^{-w} and integrating with r from 0 to 1, we have:

$$\frac{1}{2\pi i}\int_{k-i\infty}^{k+i\infty} \frac{(\frac{1+\nu+s}{2})\,\chi(s)}{(\frac{1+\nu-\alpha+s}{2})(s-w)}\,ds = G(w) = \int_{0}^{1} r^{\alpha-w}\,g(r)\,dr.$$

Let Re(w) < min. (k, $1+\alpha$). Take k' such that k' < Re(w). By the residue theory, we have, formally:

$$\frac{1}{2\pi i}\int_{k'-i\infty}^{k'+i\infty}\frac{(\frac{1+\nu+s}{2})\ \chi(s)}{(\frac{1+\nu-\alpha+s}{2})\ (s-w)}\,ds = \frac{\Gamma(\frac{1+\nu+w}{2})}{\Gamma(\frac{1+\nu-\alpha+w}{2})}\,\chi(w)$$

Now, the left-hand side of the equation is an integral which is a function analytic in the domain $Re(w) < k'$. So is the right-hand side. Thus, the function:

$$\chi(w) - \frac{G(w)\,\Gamma(\frac{1+\nu-\alpha+w}{2})}{\Gamma(\frac{1+\nu+w}{2})}$$

is analytic in the domain, $Re(w) > k'$ except at the poles of $\Gamma(\frac{1+\nu-\alpha+w}{2})$.

Consider the case $1+\nu-\alpha+k > 0$, we have, assuming the behavior at infinity:

$$\frac{1}{2\pi i}\int_{k-i\infty}^{k+i\infty}\frac{\chi(s)}{s-w}\,ds = \frac{1}{2\pi i}\int_{k-i\infty}^{k+i\infty}\frac{G(s)\Gamma(\frac{1+\nu-\alpha+s}{2})}{\Gamma(\frac{1+\nu+s}{2})(s-w)}\,ds$$

for any $Re(w) < k$. On the other hand, if k in (21) is changed into $k'(<k)$ and is then multiplied by r^{-w} where $k' < Re(w) < k$ and integrated with respect to r from 1 to ∞, we derive the following equation:

$$\frac{1}{2\pi i}\int_{k'-i\infty}^{k'+i\infty}\frac{\Gamma(\frac{1+\nu+s}{2})}{\Gamma(\frac{1+\nu+\alpha-s}{2})}\frac{\chi(s)}{s-w}\,ds = 0$$

Thus, we conclude, as before, that:

$$\frac{\Gamma(\frac{1+\nu+w}{2})}{\Gamma(\frac{1+\nu+\alpha-w}{2})}\,\chi(w)$$

is regular for $Re(w) < k$. Consequently, so is $\chi(w)$, provided that $1+\nu+\alpha-k > 0$. Assuming again the behavior of $\chi(w)$ at infinity, it follows that:

$$\frac{1}{2\pi i}\int_{k'-i\infty}^{k'+i\infty}\frac{\chi(s)}{s-w}\,ds = 0$$

$$\chi(w) = \frac{1}{2\pi i}\int_{k-i\infty}^{k+i\infty}\frac{\chi(s)}{s-w}\,ds$$

$$= \frac{1}{2\pi i}\int_{k-i\infty}^{k+i\infty}\frac{\Gamma(\frac{1+\nu-\alpha+s}{2})}{\Gamma(\frac{1+\nu+s}{2})}\frac{G(s)}{s-w}\,ds \qquad (22)$$

The last equality follows from (21). Therefore,

$$\chi(w) = \frac{1}{2\pi i}\int_{k-i\infty}^{k+i\infty}\frac{\Gamma(\frac{1+\nu-\alpha+s}{2})}{\Gamma(\frac{1+\nu+s}{2})}\,ds\int_0^1 g(\lambda)\,\lambda^{-s}\,d\lambda\int_0^1 \mu^{s-w-1}\,d\mu$$

$$= \frac{1}{2\pi i}\int_0^1 g(\lambda)\lambda^{\alpha}\,d\lambda\int_{\lambda}^1 \mu^{-w-1}d\mu\int_{k-i\infty}^{k+i\infty}\frac{\Gamma(\frac{1+\nu-\alpha+s}{2})}{\Gamma(\frac{1+\nu+s}{2})}\left(\frac{\lambda}{\mu}\right)^{-s}ds$$

$$= \frac{2}{\Gamma(\frac{1}{2}\alpha)}\int_0^1 g(\lambda)\lambda^{\alpha}d\lambda\int_0^1 \mu^{-w-1}\left(\frac{\lambda}{\mu}\right)^{\nu-\alpha+1}\left(1-\frac{\lambda^2}{\mu^2}\right)^{\frac{1}{2}\alpha-1}d\mu$$

$$= \frac{2}{\Gamma(\frac{1}{2}\alpha)}\int_0^1 \mu^{\alpha-w}\,d\mu\int_0^1 g(\rho\mu)\rho^{\nu+1}\,(1-\rho^2)^{\frac{1}{2}\alpha-1}\,d\rho \qquad (23)$$

Finally,

$$f(x) = \frac{1}{2\pi i}\int_{k-i\infty}^{k+i\infty}\frac{2^{s-\alpha}\,\Gamma(\frac{1+\nu+s}{2})}{\Gamma(\frac{1+\nu+\alpha-s}{2})}\,\chi(s)\,x^{-s}\,ds$$

$$= \frac{1}{\pi i\Gamma(\frac{1}{2}\alpha)}\int_0^1 \mu^{\alpha}d\mu\int_0^1 g(\rho,u)\,\rho^{\nu+1}\,(1-\rho^2)^{\frac{1}{2}\alpha-1}\,d\rho$$

$$\cdot\int_{k-i\infty}^{k+i\infty}2^{s-\alpha}\frac{\Gamma(\frac{1}{2}+\frac{1}{2}\nu+\frac{1}{2}s)}{\Gamma(\frac{1+\nu+\alpha-s}{2})}\,(\mu x)^{-s}\,ds \qquad (24)$$

$$= \frac{(2x)^{1-\frac{1}{2}\alpha}}{\Gamma(\frac{1}{2}\alpha)} \int_0^1 \mu^{1+\frac{1}{2}\alpha} J_{\nu+\frac{1}{2}\alpha}(\mu x) d\mu \int_0^1 g(\rho u) \rho^{\nu+1} (1-\rho^2)^{\frac{1}{2}\alpha-1} d\rho \qquad \text{(24 cont'd)}$$

The integral is meaningful for $\alpha > 0$, $\nu > -1$. This solution is obtained by Titchmarsh. For the case $\alpha < 0$, we shall derive another form of solution by the following technique:

Consider the case $\alpha > 1$, we have:

$$f(x) = \frac{(2x)^{1-\frac{1}{2}\alpha}}{\Gamma(\frac{1}{2}\alpha)} \int_0^1 \mu^{1-\frac{1}{2}\alpha-\nu} J_{\nu+\frac{1}{2}\alpha}(\mu x) \, d\mu \int_0^\mu g(\xi) \xi^{\nu+1} (\mu^2-\xi^2)^{\frac{1}{2}\alpha-1} d\xi$$

$$= \frac{(2x)^{1-\frac{1}{2}\alpha}}{\Gamma(\frac{1}{2}\alpha)} \int_0^1 g(\xi)\xi^{\nu+1} \, d\xi \int_\xi^1 \mu^{1-\frac{1}{2}\alpha-\nu} J_{\nu+\frac{1}{2}\alpha}(\mu x)(\mu^2-\xi^2)^{\frac{1}{2}\alpha-1} d\mu$$

$$= \frac{(2x)^{1-\frac{1}{2}\alpha}}{2\Gamma(\frac{1}{2}\alpha+1)} \int_0^1 g(\xi)\xi^{\nu+1} \, d\xi [J_{\nu+\frac{1}{2}\alpha}(x)(1-\xi^2)^{\frac{1}{2}\alpha} \qquad (25)$$

$$+ x \int_\xi^1 \mu^{-\frac{1}{2}\alpha-\nu} J_{\nu+\frac{1}{2}\alpha+1}(\mu x) \, (\mu^2-\xi^2)^{(\frac{1}{2})\alpha} d\mu]$$

$$= \frac{2^{-\frac{1}{2}\alpha} x^{-\alpha}}{\Gamma(\frac{1}{2}\alpha+1)} [x^{1+\frac{1}{2}\alpha} J_{\nu+\frac{1}{2}\alpha}(x) \int_0^1 g(\xi)\xi^{\nu+1} (1-\xi^2)^{\frac{1}{2}\alpha} d\xi$$

$$+ x^{2+\frac{1}{2}\alpha} \int_0^1 J_{\nu+\frac{1}{2}\alpha+1}(\mu x) \mu^{2+\frac{1}{2}\alpha} \int_0^1 g(\rho\mu)\rho^{\nu+1}(1-\rho^2)^{\frac{1}{2}\alpha} d\rho]$$

Although (25) is derived under the assumption that $\alpha > 1$, the integral is meaningful for $\alpha > -2$. One can justify that (25) is a solution of the dual integral equation provided $|\alpha-\frac{1}{2}| < 1+\nu$. The solution was first found by I.W. Busbridge as a generalization of Titchmarsh's formula by a different method. The above technique can be carried farther if g is a differentiable function.

To complete the solution of the example in (2.4), we have:

$$A(x) = \frac{f(x)}{x^2}$$

$$= \frac{2}{\pi}\frac{1}{x}\int_0^1 \frac{\rho g(\rho)d\rho}{(1-\rho^2)^{\frac{1}{2}}} + \frac{2}{\pi}\int_0^1 \frac{d\rho}{(1-\rho^2)^{\frac{1}{2}}}\int_0^1 g(\rho\mu)\sin x\mu d\mu$$

For the case $g(r) = C$ (a constant),

$$A(x) = \frac{2}{\pi} C \frac{\sin x}{x^2}$$

Thus:

$$\varphi(r,z) = \frac{2C}{\pi}\int_0^\infty \frac{\sin x}{x} e^{-xz} J_0(xr)\, dx .$$

$$= \frac{2C}{\pi} \arc\sin\left\{\frac{2}{\sqrt{(\rho-1)^2+z^2} + \sqrt{(\rho+1)^2+z^2}}\right\}$$

3.1 Applications to Problems of Elasticity

Before we are going to treat any specific problem, a brief resume is given here concerning the important features in stress analysis, strain analysis and the general stress-strain relation of homogeneous, isotropic elastic solid. The standard text should be consulted if the reader has not had the subject before. The knowledge of vector analysis is assumed. However, tensor analysis is not intended here although the components in general orthogonal coordinates are considered. Here we are concerned with the affair of two vectors. The physical and geometrical entity of stress and strain are emphasized so that relations among the components of them in general orthogonal coordinates can be established in an elementary fashion. Some of the relations are stated without the detail proof as the reader who has some elementary knowledge of elasticity and vector analysis can undoubtedly supply the proof by himself.

3.1.1. The state of stress at a point. Let $\vec{\nu}$ be any unit vector. At a point P in a medium, the force per unit area (as the area tends to zero) acting on the surface with the normal $\vec{\nu}$ from medium on the side of the normal direction is defined as the traction $\vec{T}(\vec{\nu})$ at P relative to the direction $\vec{\nu}$. The state of stress at a point is thus an affair of two vectors. Let $\vec{K}_i$ (i=1,2,3) be any three unit vectors forming a triad. Consider the equilibrium of an infinitesimal tetrahedron with the base having the (arbitrary) normal and the other edges parallel to $\vec{K}_i$ (i=1,2,3). It can be shown that $\vec{T}(\vec{\nu})$ is uniquely determined by $\vec{T}(\vec{K}_i)$ (i=1,2,3). In particular, if $\vec{K}_i$ are mutually orthogonal and if we set:

$$\vec{\nu} = \nu_i \vec{K}_i$$

and

$$\vec{T}(\vec{K}_i) = \tau_{ij} \vec{K}_j \ .$$

Where the summation convention is employed, then:

$$\vec{T}(\vec{\nu}) = \nu_\alpha \tau_{\alpha_j} \vec{K}_j \tag{26}$$

From the (moment) conditions of equilibrium, it follows that:

$$\tau_{ij} = \tau_{ji}$$

Consequently, if $\vec{K}_i{}^*$ (i=1,2,3) are any three orthogonal unit vectors such that:

$$\vec{K}_i{}^* = m_{ij} \vec{K}_j, \qquad \vec{K}_i = m_{ij}{}^* \vec{K}_j{}^* \tag{28}$$

where $m^*_{jk} = m_{kj}$ (since the transformation is a rotation and, if necessary, a reflection):

and

$$\vec{T}(K^*_i) = \tau^*_{ij} \vec{K}^*_j$$

then from (26) and (28),

$$\tau^*_{ij} \vec{K}^*_j = \tau^*_{i\alpha} m_{\alpha k} \vec{K}_k$$

$$= m_{i\alpha} \tau_{\alpha k} \vec{K}_k \tag{29}$$

Multiplying by m^*_{kj}, we have:

$$\tau^*_{ij} = m_{i\alpha}\,\tau_{\alpha k}\,m^*_{kj} = m_{i\alpha}\tau_{\alpha k}\,m_{jk} \tag{30}$$

since:

$$m_{\alpha k}\,m^*_{kj} = \delta_{\alpha j} = \begin{cases} 0, & \text{if } \alpha \neq j \\ 1, & \text{if } \alpha = j \end{cases}$$

For the sake of simplicity in presenting the results, one may write (30) in matrix notation:

$$\tau^* = M\tau M^t \tag{30'}$$

where:

$$\tau^* = (\tau^*_{ij})$$

$$\tau = (\tau_{ij})$$

$$M = (m_{ij})$$

M^t = the transpose of M.

EXERCISES

(1) The surface tractions render a body force on elementary volume. Show that, in rectangular coordinates, the equations of equilibrium are:

$$\tau_{ji,j} + F_i = \rho a_i$$

where

$$\tau_{ji,j} = \frac{\partial \tau_{ji}}{\partial x_j}$$ = the induced body force per unit volume

F_i = the ith component of actual body force, such as due to gravitation.

a_i = the ith component of the acceleration

ρ = the density of solid at the considered point.

3.1.2 The strain analysis. Let $\vec{r}$ be the position vector of the unstrained state. Let $\vec{\alpha} = \vec{r}+\vec{u}$ be that of the strained state. As long as $\vec{u}$ is given vector function of $\vec{r}$, the deformation is completely determined. At a point, what is essential to be counted for deformation is the pair of vectors $d\vec{r}$ and $d\vec{\alpha}$.

Let $d\vec{r} = a_i\vec{K}_i$ where $\vec{K}_i$ are fixed unity vectors as defined in the last section. One may write:

$$d\vec{\alpha} \cdot d\vec{\alpha} = 2\varepsilon_{ij}\, a_i\, a_j + d\vec{r} \cdot d\vec{r} \tag{31}$$

Here, $\varepsilon_{ij} = \varepsilon_{ji}$ are defined as the components of the strain tensor referring to $\vec{K}_i$. If $\vec{K}^*_i$ and $\vec{K}_i$ are related as in the last section, we have:

$$d\vec{r} = a^*_i\, \vec{K}^*_i = a^*_j \cdot m_{ji}\, \vec{K}_i$$

or:

$$a_i = m_{ji}\, a^*_j \tag{32}$$

Now both $d\vec{\alpha} \cdot d\vec{\alpha}$ and $d\vec{r} \cdot d\vec{r}$ are invariant under orthogonal transformation, hence:

$$\varepsilon^*_{ij}\, a^*_i\, a^*_j = \varepsilon_{ij}\, a_i\, a_j\ .$$

Thus:

$$\varepsilon^*_{ij} = m_{ik}\, \varepsilon_{kl}\, m_{jl}\ . \tag{33}$$

In the matrix notation, we have:

$$\varepsilon^* = M\varepsilon M^t \tag{34}$$

3.1.3 Stress and strain relation. The matrix of components of stress or strain referring to fixed orthogonal directions at a point is symmetrical and, hence it can also always be reduced to a diagonal matrix by the change of coordinates. These directions, to which the shear components of the stress (or strain) vanish, are called principal axis for stress (or strain). The following assumptions are imposed:

(i) The material is homogeneous and isotropic.

(ii) The state of stress is uniquely and continuously determined by that of strain. Hence, the physical law is independent of choice coordinates.

Referring to $\vec{K}_i (i=1,2,3)$ (ii) states that τ_{ij} are definite functions of all the components of ε_{lk}. We may write it in the matrix form:

$$\tau = F(\varepsilon) \tag{35}$$

where F is a matrix. Referring to $\vec{K}_i^*$, we have also:

$$\tau^* = F(\varepsilon^*) \tag{36}$$

From (30') and (34), it follows that:

$$F(\varepsilon) = M^t F(M\varepsilon M^t)M$$

where $M^t M = I$ = the identity matrix.

Thus, for arbitrary symmetric ε, F has the following invariant properties:

$$MF(\varepsilon)M^t = F(M\varepsilon M^t) \tag{37}$$

where M is any orthogonal matrix.

Thus, without loss of generality, we may assume that:

$$\varepsilon^* = \begin{pmatrix} \lambda_1 & 0 & 0 \\ 0 & \lambda_2 & 0 \\ 0 & 0 & \lambda_3 \end{pmatrix} \tag{38}$$

By the virtue of the assumption (ii),

$$\tau^* = \begin{pmatrix} \mu_1 & 0 & 0 \\ 0 & \mu_2 & 0 \\ 0 & 0 & \mu_3 \end{pmatrix}$$

Now,

$$\mu_i = f_i(\lambda_1, \lambda_2, \lambda_3) \qquad (i = 1,2,3) \tag{39}$$

From (37), for any orthogonal matrix P, we have

$$P\tau^* p^t = F(P\varepsilon^* p^t) \tag{40}$$

In particular, if we take:

$$P = \begin{pmatrix} 0, & 1, & 0 \\ 1, & 0, & 0 \\ 0, & 0, & 1 \end{pmatrix}$$

the left-hand side of (40) is simply the matrix obtained by interchanging the first and the second rows and columns. Hence:

$$f_3(\lambda_2, \lambda_1, \lambda_3) = f_3(\lambda_1, \lambda_2, \lambda_3) \tag{41}$$

$$f_2(\lambda_1, \lambda_2, \lambda_3) = f_1(\lambda_2, \lambda_1, \lambda_3) \tag{42}$$

(41) means that f_3 is a symmetric function of λ_1, λ_2. If now, one writes:

$$f_3(x; \lambda_1\ \lambda_2\ \lambda_3) = [f_3(\lambda_1,\lambda_2,x) + f_3(\lambda_1,\lambda_3,x) + f_3(\lambda_3,\lambda_2,x)]$$

$$- [f_3(\lambda_1,x,x) + f_3(\lambda_2,x,x) + f_3(\lambda_3,x,x)] + f_3(x,x,x)$$

we have:

$$F_3(\lambda_3;\lambda_1,\lambda_2,\lambda_3) = f_3(\lambda_1,\lambda_2,\lambda_3)$$

It is clear that F_3 is a function of x and is symmetric with respect to the three parameters λ_1, λ_2, λ_3. One may also write:

$$F_3(x,\lambda_1,\lambda_2,\lambda_3) = F_3^*(x; a,b,c)$$

where:

$$a = \lambda_1 + \lambda_2 + \lambda_3$$

$$b = \lambda_1 \cdot \lambda_2 + \lambda_2 \cdot \lambda_3 + \lambda_3, \lambda_1$$

$$c = \lambda_1 \cdot \lambda_2 \cdot \lambda_3$$

Consequently, one can also write:

$$F_i^* (\lambda_i, a, b, c) = F_i(\lambda_i; \lambda_1, \lambda_2, \lambda_3) \qquad (i = 1,2,3)$$

From (42):

$$F_2^*(\lambda_2; a, b, c) = F_1^* (\lambda_2; a, b, c)$$

Hence, we can finally put:

$$F_i^* (x; a, b, c) = F^*(x; a, b, c) \qquad (i = 1,2,3) \qquad (43)$$

Therefore:

$$\tau^* = \begin{vmatrix} F^*(\lambda_1; a, b, c) & 0, & 0 \\ 0, & F^*(\lambda_2; a, b, c) & 0 \\ 0, & 0, & F^*(\lambda_3; a, b, c) \end{vmatrix}$$

Here, we are only concerning the value of the function F^* at λ_1, λ_2, λ_3. Thus, it can be replaced by its interpolation polynomial (which is uniquely defined by the values of F^* at λ_i in the case that λ_i are distinct). Thus, in the case of distinct λ_i, we have the polynomial.

$$Q(x) = g_o(a,b,c) + g_1(a,b,c)x + g_2(a,b,c)x^2$$

such that

$$Q(\lambda_i) = F(\lambda_i;a,b,c) \qquad (i = 1,2,3) .$$

As a matter of fact, one has explicitly:

$$Q(x) = \frac{\begin{vmatrix} 0 & 1 & x & x^2 \\ F(\lambda_1) & 1 & \lambda_1 & \lambda_1^2 \\ F(\lambda_2) & 1 & \lambda_2 & \lambda_2^2 \\ F(\lambda_3) & 1 & \lambda_3 & \lambda_3^2 \end{vmatrix}}{-\begin{vmatrix} 1 & 1 & \lambda_1^2 \\ 1 & \lambda_2 & \lambda_2^2 \\ 1 & \lambda_3 & \lambda_3^2 \end{vmatrix}} \tag{44}$$

Consequently, we have:

$$\tau^* = \begin{pmatrix} Q(\lambda_1) & 0 & 0 \\ 0 & Q(\lambda_2) & 0 \\ 0 & 0 & Q(\lambda_3) \end{pmatrix} = g_0(a,b,c)I + g_1(a,b,c)\varepsilon^* + g_2\varepsilon^{*2} \tag{45}$$

Or, in general, *

$$\tau = g_0(a,b,c)I + g_1(a,b,c)\varepsilon + g_2(a,b,c)\varepsilon^2 \tag{46}$$

The expression (46) represents a general Stress-Strain Law. It is clear that if a further assumption is imposed that the relation is linear and homogeneous, we derive immediately Hooke's Law:

$$\tau = \lambda a I + 2\mu\varepsilon \tag{47}$$

---

*This general form was pointed out independently by W. Prager (J. Appl. Phys. 16, p. 837, 1945, and Reiner (Am. J. Math. 67, p. 350, 1945) with the assumption that τ_{ij} are polynomial of ε_{kl}. The present derivation drops this restriction. See, also, H. Richter, uber Matrixfunctionen, Math. Ann. Vol. 122, pp. 16-34, (1950). For the derivation of the law based on the principle of virtual displacements, see for instance, Novozhilov, V.V., Foundation of Nonlinear Theory of Elasticity, (English translation by Graylock), pp. 101-112, 1953.

where λ and 2μ are (Lame's) constants. They are related with commonly used elastic constant, Young's modulus E and Poisson's ratio σ:

$$E = \frac{\mu(3\lambda+2\mu)}{\lambda+\mu} \tag{48}$$

$$\sigma = \frac{\lambda}{2(\lambda+\mu)} \tag{49}$$

EXERCISE 10

(1) Stoke's Law for Viscous Fluids

(i) Analogous to Hooke's Law, let us assume that the stress tensor be decomposed into:

$$-pI + \tau' .$$

where the static pressure defined as the average of the normal stresses:

$$-p = \frac{\tau_{11}+\tau_{22}+\tau_{33}}{3}$$

and τ' is the stress due to the motion of fluid. Instead of stress-strain relation, we assume the stress and strain rate relation.

Thus:

$$\tau' = \lambda(\varepsilon'_{kk})I + 2\mu\varepsilon'$$

where:

$$\varepsilon'_{ik} = \frac{d}{dt}\varepsilon_{ik}$$

Show that by the definition of the static pressure

$$\lambda = -\frac{2\mu}{3}$$

(ii) Derive the equation of motion (Stokes-Navier's equations) in vector form:

$$\rho\frac{D\vec{q}}{Dt} = \vec{G} + \frac{\mu}{3}\ (\text{div. } \vec{q}) - \nabla p + \mu\nabla^2\vec{q}$$

where $\vec{G}$ is the body force for per unit volume.

$$\frac{D}{Dt} = \frac{\partial}{\partial t} + \vec{q}\cdot\nabla$$

or, in rectangular coordinates,

$$\rho \frac{Du}{Dt} = G_{(x)} + \frac{\mu}{3}\frac{\partial}{\partial x}\left(\frac{\partial u}{\partial x} + \frac{\partial v}{\partial y} + \frac{\partial w}{\partial z}\right) - \frac{\partial}{\partial x} p$$

$$+ \mu\left(\frac{\partial^2}{\partial x^2} + \frac{\partial^2}{\partial y^2} + \frac{\partial^2}{\partial z^2}\right) u$$

etc., where $\vec{q} = u\vec{k}_{(x)}, + v\vec{k}_{(y)}, + w\vec{k}_{(z)}, \vec{k}_{(x)}$ = the unit vector in the x-direction and:

$$\frac{Du}{Dt} = \frac{\partial u}{\partial t} + u\frac{\partial u}{\partial x} + v\frac{\partial u}{\partial y} + w\frac{\partial u}{\partial z}$$

<u>3.1.4 Orthogonal coordinates</u>. It is often desirable to employ general orthogonal coordinates for a problem with a certain specific boundary conditions (such as due to a specific geometrical shape of the physical boundary). In order to calculate the strain components and to transform the equilibrium equations in the general orthogonal coordinates, the following results concerning the rotation of the local frame of references due to the change of point to point are first derived. Let the orthogonal coordinates be defined by:

$$\vec{x} = \vec{f}(q_1, q_2, q_3)$$

where $\vec{x}$ is the position vector of the point $P(q_1, q_2, q_3)$, we shall define the q_i-direction as that along the increment of q_i. The unit vector parallel to the q_i-direction is denoted by $\vec{K}_i$. It is clear that $\vec{K}_i$ depends on the point $P(q_1, q_2, q_3)$. When P is changed into $Q(q_1', q_2', q_3')$ where $q_i' = q_i + dq_i$, the triad composed of $\vec{K}_1$, $\vec{K}_2$ and $\vec{K}_3$ is subjected to a rotation. Let us designate this rotation by $d\omega = d\omega_j\ K_j(q_1, q_2, q_3)$. First of all, we would like to find these components $d\omega_i$. Let $d\vec{x} = dx_j K_j(q_1, q_2, q_3)$. Hence:

$$ds^2 = dx_k \cdot dx_k = H_{ij} dq_i dq_j$$

For orthogonal coordinates, if $i \neq j$

$$H_{ij} = 0 .$$

Thus, we put

$$H_{ij} = \delta_{ik}\delta_{j\ell}h_k h_\ell \quad ,$$

where:

$$h_\ell (\ell = 1,2,3)$$

are positive-valued functions of q_1, q_2 and q_3. Now, consider an increment dq_1 of q_1. This introduces a special rotation

$$d\vec{\omega}^{(1)} = d\vec{\omega}_j^{(1)}\vec{K}_j(q_1, q_2, q_3)$$

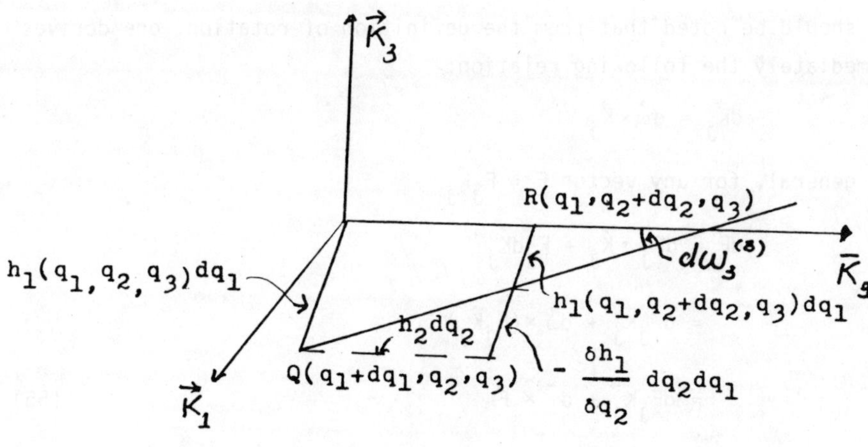

Figure 11

Referring to the Figure 1, we have, neglecting infinitesimals of higher order,

$$d\omega_3^{(1)} = \sin d\omega_3^{(1)} = -\frac{1}{h_2}\frac{\partial h_1}{\partial q_2}dq_1 \tag{50}$$

Similarly:

$$d\omega_2^{(1)} = \frac{1}{h_3}\frac{\partial h_1}{\partial q_3}\,dq_1 \tag{51}$$

Thus, in general,* we have:

$$d\vec{\omega} = d\vec{\omega}^{(1)} + d\vec{\omega}^{(2)} + d\vec{\omega}^{(3)} = d\omega_j\vec{K}_j(q_1, q_2, q_3)$$

or

$$d\omega_1 = \frac{1}{h_2}\frac{\partial h_3}{\partial q_2}\,dq_3 - \frac{1}{h_3}\frac{\partial h_2}{\partial q_3}\,dq_2 \tag{52}$$

$$d\omega_2 = \frac{1}{h_3}\frac{\partial h_1}{\partial q_3}\,dq_1 - \frac{1}{h_1}\frac{\partial h_3}{\partial q_1}\,dq_3 \tag{53}$$

$$d\omega_3 = \frac{1}{h_1}\frac{\partial h_2}{\partial q_1}\,dq_2 - \frac{1}{h_2}\frac{\partial h_1}{\partial q_2}\,dq_1 \tag{54}$$

It should be noted that from the definition of rotation, one derives immediately the following relation:

$$d\vec{K}_j = d\vec{\omega}\times\vec{K}_j$$

In general, for any vector $F = F_j\vec{K}_j$

$$\begin{aligned} d\vec{F} &= dF_j\cdot\vec{K}_j + F_j d\vec{K}_j \\ &= dF_j\vec{K}_j + d\vec{\omega}\times(F_j\vec{K}_j) \\ &= dF_jK_j + d\vec{\omega}\times\vec{F} \end{aligned} \tag{55}$$

Thus, explicitly we have:

$$\begin{aligned} d\vec{K}_1 &= d\vec{\omega}\times\vec{K}_1 = -\,d\omega_2\vec{K}_3 + d\omega_3\vec{K}_2 \\ &= \left(\frac{1}{h_1}\frac{\partial h_2}{\partial q_1}\,dq_2 - \frac{1}{h_2}\frac{\partial h_1}{\partial q_2}\,dq_1\right)\vec{K}_2 + \left(\frac{1}{h_1}\frac{\partial h_3}{\partial q_1}\,dq_3 - \frac{1}{h_3}\frac{\partial h_1}{\partial q_3}\,dq_3\right)\vec{K}_3 \end{aligned} \tag{56a}$$

---

*For infinitesimal rotations, the order of applications is immaterial and the superposition principle applies.

$$d\vec{K}_2 = (\frac{1}{h_2}\frac{\partial h_1}{\partial q_2}\, dq_1 - \frac{1}{h_1}\frac{\partial h_2}{\partial q_1}\, dq_2)\vec{K}_1 + (\frac{1}{h^2}\frac{\partial h_3}{\partial q_2}\, dq_3 - \frac{1}{h_3}\frac{\partial h_2}{\partial q_3}\, dq_2)\vec{K}_3 \tag{56b}$$

$$d\vec{K}_3 = (\frac{1}{h_3}\frac{\partial h_1}{\partial q_3}\, dq_1 - \frac{1}{h_1}\frac{\partial h_3}{\partial q_1}\, dq_3)\vec{K}_1 + (\frac{1}{h_3}\frac{\partial h_2}{\partial q_3}\, dq_2 - \frac{1}{h^2}\frac{\partial h_3}{\partial q_2}\, dq_3)\vec{K}_2 \tag{56c}$$

Or, we write

$$d\vec{K}_i = e_{iks}\, dq_s \vec{K}_k$$

where e_{iks} are well-defined by (56a), (56b) and (56c).

3.1.5 Equations of equilibrium in orthogonal coordinates. Consider a point $P(q_1, q_2, q_3)$ in the strained state and a small volume element as indicated in the Figure 12.

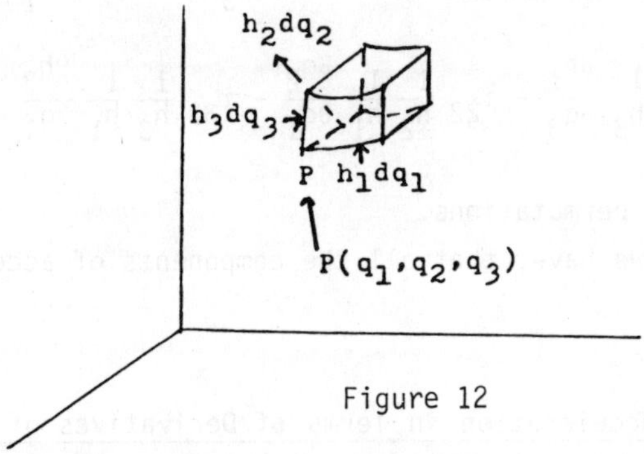

Figure 12

The surface force acting on the elementary volume induces a body force (per unit volume). The contribution due to the resultant from tractions on surface corresponding to q_1 and q_1+dq_1 is, except infinitesimals of higher order:

$$\frac{dq_2 dq_3}{h_1 h_2 h_3 dq_1 dq_2 dq_3}\left[h_2 h_3 \tau_{1j}\vec{K}_i\Big|_{q_1+dq_1} - h_2 h_3 \tau_{1j}\vec{K}_i\Big|_{q_1}\right]$$

$$= \{\frac{1}{h_2 h_3}\frac{1}{h_1}\frac{\partial}{\partial q_1}\ \tau_{1j} h_2 h_3)\}\ \vec{K}_i + \frac{1}{h_1}\tau_{1j} e_{jk1}\, dq_1 \vec{K}_k$$

Thus, the total induced body force (per unit volume) is:

$$[\frac{1}{h_1h_2h_3}\frac{\partial}{\partial q_1}(h_2h_3\tau_{1j}) + \frac{1}{h_1h_2h_3}\frac{\partial}{\partial q_2}(h_3h_1\tau_{2j}) + \frac{1}{h_1h_2h_3}\frac{\partial}{\partial q_3}(h_1h_2\tau_{3j})$$

$$+ \tau_{ij}\frac{1}{h_1}e_{ji1} + \tau_{2_j}\frac{1}{h_2}e_{ji2} + \tau_{3_j}\frac{1}{h_3}e_{ji3}]\,\vec{K}_i\cdot = B_iK_i\ .$$

Let ρ be the density, $\vec{a} = a_i\vec{K}_i$ the acceleration vector and $F = F_i\vec{K}_i$ the external body force (per unit volume), we have the equilibrium equations:

$$\rho a_i = F_i + B_i$$

or, explicitly, we have:

$$\rho a_1 = \frac{1}{h_1h_2h_3}[\frac{\partial h_2h_3\tau_{11}}{\partial q_1} + \frac{\partial h_3h_1\tau_{21}}{\partial q_2} + \frac{\partial h_1h_2\tau_{31}}{\partial q_3}] + \tau_{12}\frac{1}{h_1}\frac{1}{h_2}\frac{\partial h_1}{\partial q_2}]$$

$$+ \tau_{13}\frac{1}{h_1}\frac{1}{h_3}\frac{\partial h_1}{\partial q_3} - \tau_{22}\frac{1}{h_2}\frac{1}{h_1}\frac{\partial q_2}{\partial q_1} - \tau_{33}\frac{1}{h_3}\frac{1}{h_1}\frac{\partial h_3}{\partial q_1}$$

and others by cyclic permutations.
In the static case, we have, that all the components of acceleration vanish.

EXERCISE:

Components of Acceleration in Terms of Derivatives of those of Displacement in Orthogonal Coordinates.

Let $p(q_1, q_2, q_3)$ be a point of the medium in strained state. Let $\vec{u}(q_1, q_2, q_3, t)$ be the displacement vector from unstrained state to the strained state at the instant t. Following the same small element of volume, the orthogonal coordinates q_i are considered as functions of t. Thus, we have the velocity-vector:

$$\vec{V} = \frac{du_i}{dt}\vec{K}_i(q_1, q_2, q_3)$$

where

$$\frac{du}{dt} = \frac{\partial u}{\partial t} + \dot{q}_j \frac{\partial u}{\partial q_j}$$

On the other hand, we have:

$$\vec{V} = \delta_{ij} h_i \dot{q}_i \vec{K}_j$$

Thus:

$$\dot{q}_1 = \frac{1}{h_1} \frac{du_1}{dt} \text{ etc.}$$

Consequently,

$$\frac{\partial u_1}{\partial t} + \frac{1}{h_1} V_1 \frac{\partial u_1}{\partial q_1} + \frac{1}{h_2} V_2 \frac{\partial u_1}{\partial q_2} + \frac{1}{h_3} V_3 \frac{\partial u_1}{\partial q_3} = V_1$$

$$\frac{\partial u_2}{\partial t} + \frac{1}{h_1} V_1 \frac{\partial u_2}{\partial q_1} + \frac{1}{h_2} V_2 \frac{\partial q_2}{\partial q_2} + \frac{\partial u_3}{\partial q_3} = V_2$$

$$\frac{\partial u_3}{\partial t} + \frac{1}{h_1} V_1 \frac{\partial u_3}{\partial q_1} + \frac{1}{h_2} V_2 \frac{\partial u_3}{\partial q_2} + \frac{1}{h_3} V_3 \frac{\partial u_3}{\partial q_3} = V_3$$

When

$$\delta_{jk} \frac{1}{h_j} \frac{\partial u_i}{\partial q_j} \qquad (i,k = 1,2,3)$$

are small, then:

$$V_i = \frac{\partial u_i}{\partial t} .$$

Find the velocity-components as well as the acceleration components in cylindrical coordinates in terms of the derivatives of the components of displacement with respect to the coordinates and time.

<u>HINT:</u> Show that,

$$a_1 = \frac{dV_1}{dt} + V_2 \left(\frac{1}{h_2} \frac{\partial h_1}{\partial q_2} \frac{V_1}{h_1} - \frac{1}{h_1} \frac{\partial h_2}{\partial q_1} \frac{V_2}{h_2}\right)$$
$$+ V_3 \left(\frac{1}{h_3} \frac{\partial h_1}{\partial q_3} \frac{V_1}{h_1} - \frac{1}{h_1} \frac{\partial h_3}{\partial q_1} \frac{V_3}{h_3}\right) \text{, etc.}$$

If (u,v,w) are components of the displacement vector in dr, dθ and dz directions at (r,θ,z) respectively, then we have:

$$V_{(r)} = \frac{-1}{D}\begin{vmatrix} u_t, & \frac{1}{r}u_\theta, & u_z \\ v_t, & \frac{1}{r}v_\theta - 1, & v_z \\ w_t, & \frac{1}{r}w_\theta, & w_z - 1 \end{vmatrix} \quad (57a)$$

$$V_\theta = \frac{-1}{D}\begin{vmatrix} u_r - 1, & u_t, & u_z \\ v_r, & v_t, & v_z \\ w_r, & w_t, & w_z - 1 \end{vmatrix} \quad (57b)$$

$$V_{(z)} = \frac{-1}{D}\begin{vmatrix} u_r - 1, & \frac{1}{r}u_\theta, & u_t \\ v_r, & \frac{1}{r}v_\theta - 1, & v_t \\ w_r, & \frac{1}{r}w_\theta, & w_t \end{vmatrix} \quad (57c)$$

where:

$$D = \begin{vmatrix} u_r - 1, & \frac{1}{r}u_\theta, & u_z \\ v_r, & \frac{1}{r}v_\theta - 1, & v_z \\ w_r, & \frac{1}{r}w_\theta, & w_z - 1 \end{vmatrix} \quad (58)$$

<u>ANSWER</u>.

$$a_{(r)} = \frac{d}{dt}\left(\frac{du}{dt}\right) - \frac{V_{(\theta)}^2}{r}$$

$$a_{(\theta)} = \frac{d}{dt}\left(\frac{dv}{dt}\right) + \frac{V_{(r)}V_{(\theta)}}{r}.$$

and

$$a_{(z)} = \frac{d}{dt}\left(\frac{dw}{dt}\right)$$

3.1.6 Components of strain tensor in orthogonal coordinates. As the fundamental components of stress tensor are referred to the local frame of references at the point $P(q_1, q_2, q_3)$ in strained state, we have also to find the fundamental components of strain referring to the same. We shall also consider the displacement vector as the vector function of the point $P(q_1, q_2, q_3)$ of the strained state. Let

$$d\vec{r} = h_i(q_1, q_2, q_3)\, dq_i \vec{K}_i$$

and

$$d\vec{\alpha} = h_i(q_1, q_2, q_3)\, dq_i^* \vec{K}_i$$

where:

$$h_1 dq_1^* = h_1 dq_1 + u_{1,j}\, dq_j^* + e_{21s}\, dq_s^*\, u_2$$

$$+ e_{31s}\, dq_s^*\, u_3\,,$$

where

$$u_{i,j} = \frac{\partial u_i}{\partial q_j}$$

and the other two relations by cyclic permutations e_{ijk} are defined by (56). Or, explicitly, we have:

$$h_1 dq_1 = \left(1 - \frac{1}{h_1}\frac{\partial u_1}{\partial q_1} - \frac{u_2}{h_1}\frac{\partial h_1}{h_2 \partial q_2} - \frac{u_3}{h_1}\frac{1}{h_3}\frac{\partial h_1}{\partial q_3}\right) h_1 dq_1^*$$

$$+ \left(- \frac{1}{h_2}\frac{\partial u_1}{\partial q_2} + \frac{1}{h_2}\frac{u_2}{h_1}\frac{\partial h_2}{\partial q_1}\right) h_2 dq_2^*$$

$$+ \left(- \frac{1}{h_3}\frac{u_1}{q_3} + \frac{1}{h_3}\frac{u_3}{h_1}\frac{h_3}{q_1}\right) h_3 dq_3^* \quad \text{etc.} \tag{59}$$

The relation (59) can be put into matrix form:

$$\begin{bmatrix} h_1 dq_1 \\ h_2 dq_2 \\ h_3 dq_3 \end{bmatrix} = (a_{ij}) \begin{bmatrix} h_1 dq_1^* \\ h_2 dq_2^* \\ h_3 dq_3^* \end{bmatrix} \tag{60}$$

Let:

$$A = (a_{ij})$$

For small displacements and small deformations, A is an identity matrix. Now, set:

$$2\varepsilon_{ij}(h_i dq_i)(h_j dq_j) = 2\varepsilon^*_{ij}(h_i dq^*_i)(h_j dq^*_j)$$

$$\varepsilon^* = A\varepsilon A^t. \qquad (61)$$

or:

$$\varepsilon = A^{-1}\varepsilon^*(A^t)^{-1} \qquad (61)$$

The elements of the matrix ε^* are explicitly expressed in terms of the derivatives of the components of the displacement; namely, we have:

$$2\varepsilon^*_{ij}h_i h_j dq_i dq_j = d\vec{\alpha}\cdot d\vec{\alpha} - d\vec{r}\cdot d\vec{r}$$

$$= h_i^2 dq_i^{*2} - h_i^2 dq_i^2$$

and (59).

<u>EXERCISE:</u>

Show that when both the deformation and the displacement are small, we have, by neglecting the quadratic terms of u_i and their derivatives:

$$\varepsilon^*_{ij} = \varepsilon_{ij}$$

and:

$$\varepsilon^*_{11} = \frac{1}{h_1}\frac{\partial u_1}{\partial q_1} + \frac{u_2}{h_1}\frac{1}{h_2}\frac{\partial h_1}{\partial q_2} + \frac{u_3}{h_1}\frac{1}{h_3}\frac{\partial h_1}{\partial q_3} \qquad (62a)$$

$$\varepsilon^*_{22} = \frac{1}{h_2}\frac{\partial u_2}{\partial q_2} + \frac{u_3}{h_2}\frac{1}{h_3}\frac{\partial h_2}{\partial q_3} + \frac{u_1}{h_2}\frac{1}{h_1}\frac{\partial h_2}{\partial q_1} \qquad (62b)$$

$$\varepsilon^*_{33} = \frac{1}{h_3}\frac{\partial u_3}{\partial q_3} + \frac{\partial u_1}{h_3}\frac{1}{h_1}\frac{\partial h_3}{\partial q_1} + \frac{u_2}{h_3}\frac{1}{h_2}\frac{\partial h_3}{\partial q_2} \qquad (62c)$$

$$\varepsilon^*_{12} = \varepsilon^*_{21} = \tfrac{1}{2}\left[\frac{h_1}{h_2}\frac{\partial}{\partial q_2}\left(\frac{u_1}{h_1}\right) + \frac{h_2}{h_1}\frac{\partial}{\partial q_1}\left(\frac{u_2}{h_2}\right)\right] \tag{62d}$$

$$\varepsilon^*_{13} = \varepsilon^*_{31} = \tfrac{1}{2}\left[\frac{h_1}{h_3}\frac{\partial}{\partial q_3}\left(\frac{u_1}{h_1}\right) + \frac{h_3}{h_1}\frac{\partial}{\partial q_1}\left(\frac{u_3}{h_3}\right)\right] \tag{62e}$$

$$\varepsilon^*_{23} = \varepsilon^*_{32} = \tfrac{1}{2}\left[\frac{h_2}{h_3}\frac{\partial}{\partial q_3}\left(\frac{u_2}{h_2}\right) + \frac{h_3}{h_2}\frac{\partial}{\partial q_2}\left(\frac{u_3}{h_3}\right)\right] \tag{62f}$$

<u>3.1.7 Linear static problems with axially symmetrical stress distribution</u>. By means of Section 3.1.5, we have, for the case of rotational symmetry without external body force, the equations of equilibrium in cylindrical coordinates:

$$\frac{\partial\tau_{(rr)}}{\partial r} + \frac{\partial\tau_{(rz)}}{\partial z} + \frac{\tau_{(rr)} - \tau_{(\theta\theta)}}{r} = 0 \tag{63}$$

and

$$\frac{\partial\tau_{(rz)}}{\partial r} + \frac{\partial\tau_{(zz)}}{\partial z} + \frac{\tau_{(rz)}}{r} = 0 \tag{64}$$

since (i) none of the stress components depends on θ, and (ii)

$$\tau_{(r\theta)} = \tau_{(z\theta)} = 0 .$$

From Hooke's stress and strain law and Ex. 1 of the last section, it follows that:

(1) $u = u(r,z)$, $v = 0$, $w = w(r,z)$;

and

(2) $$\tau_{(rr)} = \lambda(u_r + \frac{1}{r}v_\theta + \frac{1}{r}u + w_z) + 2\mu u_r \tag{65a}$$

$$= (\lambda+2\mu)u_r + \lambda\frac{u}{r} + \lambda w_z$$

$$\tau_{(\theta\theta)} = \lambda u_r + (\lambda+2\mu)\frac{u}{r} + \lambda w_z \tag{65b}$$

$$\tau_{(zz)} = \lambda(u_r + \frac{1}{r}u) + (\lambda+2\mu)\, w_z \tag{65c}$$

$$\tau_{(rz)} = \mu(u_z+w_r) \tag{65d}$$

where

$$u_r = \frac{\partial u}{\partial r}$$

and so on.

Substituting (63) and (64), it yields the following equations in u and w:

$$(\lambda+2\mu)\frac{\partial}{\partial r}\left(u_r + \frac{1}{r}u\right) + \mu u_{zz} + (\lambda+\mu)\, w_{rz} = 0 \tag{66}$$

$$(\lambda+\mu)\frac{\partial}{\partial z}\left(u_r + \frac{1}{r}u\right) + \mu\Delta w + (\lambda+\mu)\, w_{zz} = 0 \tag{67}$$

where:

$$\Delta = \frac{\partial^2}{\partial r^2} + \frac{1}{r}\frac{\partial}{\partial r} + \frac{\partial^2}{\partial z^2}$$

If, we set:

$$u = -\frac{\lambda+\mu}{\mu}\,\Phi_{rz} \tag{68}$$

which is suggested by (66), the same equation immediately yields:

$$w = \frac{\lambda+2\mu}{\mu}\Delta\Phi - \frac{\lambda+\mu}{\mu}\Phi_{zz} \tag{69}$$

Substituting both (68) and (69) in (67), we have:

$$\Delta\Delta\Phi = 0 \tag{70}$$

or:

$$\left(\frac{\partial^2}{\partial r^2} + \frac{1}{r}\frac{\partial}{\partial r} + \frac{\partial^2}{\partial z^2}\right)^2 \Phi = 0 \tag{70'}$$

Let $\Phi^*(\xi,z)$ be the Hankel's transform of $\Phi(r,z)$: namely,

$$\Phi^*(\xi,z) = \int_0^\infty \Phi(r,z)\, r J_0(r,\xi)\, dr \tag{71}$$

Thus, the Hankel transform of (70) is:

$$\left(\frac{\partial^2}{\partial z^2} - \xi^2\right)^2 \Phi^* = 0$$

$$\text{or } \Phi^*(\xi,z) = (A+Bz)e^{-\xi z} + (C+Dz)e^{\xi z} \tag{72}$$

EXERCISE:

Show that:

$$u = -\frac{\lambda+\mu}{\mu}\frac{\partial^2}{\partial r\partial z}\int_0^\infty \Phi^*(\xi,z)\xi J_0(r\,\xi)\,d\xi$$

$$= \frac{\lambda+\mu}{\mu}\frac{\partial}{\partial z}\int_0^\infty \xi^2\Phi^*(\xi,z)J_1(r\,\xi)\,d\xi \tag{73a}$$

$$w = \int_0^\infty \xi\left(\Phi^*_{zz} - \frac{\lambda+2\mu}{\mu}\xi^2\Phi^*\right) J_0(\xi\,r)\,d\xi \tag{73b}$$

$$\tau_{(zz)} = \int_0^\infty \xi[(\lambda+2\mu)\Phi^*_{zzz} - (3\lambda+4\mu)\xi^2\Phi^*]\,J_0(\xi\,r)\,d\xi \tag{74a}$$

$$\tau(r\,z) = \int_0^\infty \xi^2[\lambda\Phi^*_{zz} + (\lambda+2\mu)\xi^2\Phi^*]J_1(\xi\,r)\,d\xi \tag{74b}$$

$$\tau_{(r\,r)} = \int_0^\infty \xi[\lambda\Phi^*_{zzz} + (\lambda+2\mu)\xi^2\Phi^*_z]\,J_0(\xi\,r)\,d\xi$$

$$-\frac{2(\lambda+\mu)}{r}\int_0^\infty \xi^2\Phi^*_z\,J_1(\xi\,r)\,d\xi \tag{74c}$$

$$\tau_{(\theta\theta)} = \lambda\int_0^\infty \xi[\Phi^*_{zzz}-\xi^2\Phi^*]J_0(\xi\,r)\,d\xi + \frac{2(\lambda+\mu)}{r}\int_0^\infty \xi^2\Phi^*_z J_1(\xi\,r)\,d\xi \tag{74d}$$

3.1.7.1 Indentation of the plane surface of semi-infinite elastic medium by a rigid punch.* We have, now the following boundary-value problem:

*J.W. Harding and I.N. Sniddon, Proc. Cambridge Phil. Soc., Vol. 41, p. 16, (1945).

$$\Delta^2\Phi = 0 \tag{75}$$

where:

$$\Delta = \frac{\partial^2}{\partial r^2} + \frac{1}{r}\frac{\partial}{\partial r} + \frac{\partial^2}{\partial z^2} \quad .$$

The boundary conditions are:

$$w(r,0) = f(r) \;, \qquad (0 \le r \le a) \tag{76a}$$

$$\tau_{(zz)}(r,0) = 0 \qquad (r > a) \tag{76b}$$

$$\tau_{(rz)}(r,0) = 0 \qquad (0 < r < \infty) \tag{77}$$

where w, $\tau_{(zz)}$, and $\tau_{(rz)}$, are given by (65) and (68) and (69). On the other hand, all the stress components and displacement components tend to zero as z tends to infinity. From (72), it follows that C = D = 0. By means of (77), we have:

$$(\lambda+\mu)\xi A = \lambda B$$

or:

$$\Phi^*(\xi,z) = B(\xi)\left[\frac{\lambda\xi}{\lambda+\mu} + z\right] e^{-\xi z}$$

Then the conditions (76a) and (76b) yield:

$$\int_0^\infty \xi^2 B(\xi) J_0(\xi\, r)\, d\xi = \frac{-\mu}{\lambda+\mu} f(r) \;, \qquad (0 < r \le a) \tag{78a}$$

and:

$$\int_0^\infty \xi^3 B(\xi) J_0(\xi\, r)\, d\xi = 0 \;, \qquad (r > a) \tag{78b}$$

CHAPTER V.

INTRODUCTION TO POTENTIAL THEORY

The first part of the present chapter deals with the boundary-value problems of the Laplace equation. The simplest case of the circle problem is solved by means of Fourier series. The formulas of Schwarz, Poisson, Boggio and Dini for the circle problem are derived. The method of conformal mapping is the introduced in attacking the general two-dimensional problems. In attacking higher dimensional problems, the notion of Green's function and the method of images are used. Extension of these methods to other types of partial differential equations is illustrated by examples, such as the problems of vibrating airfoils in a supersonic flow and of the bending of an elastic plate under a concentrated load. The methods of a simple layer and a double layer and some important theorems concerning harmonic functions are included in exercises.

1. THE BOUNDARY-VALUE PROBLEMS OF LAPLACE EQUATIONS

1.1 Interior Problems

In dealing with the problems such as to determine the two-dimensional steady temperature distribution in an infinite cylinder with the steady boundary control on the lateral surface the following types of mathematical problems are confronted:

<u>1.1.1 The Dirichlet problem</u>. The boundary value of the temperature is prescribed. Let D be the two-dimensional domain which, at the present, is the cross section of the cylinder. Let D be bounded by a finite number of non-intersecting closed (Jordan) curves* B_j ($j = 1,2,...,m$). If θ is the parameter for the parametric representation of the boundary $B = B_1 + B_2 + ... + B_m$, then the problem is:

---

*For the interior problems, one of the boundary curves B_j contains all the others if any, in its interior. While for the exterior problems, each B_j lies outside of the rest.

$$\Delta u = 0 . \tag{1}$$

with the boundary value on B prescribed as

$$u = f(\theta) \ . \tag{2}$$

One can, of course, interpret u as the electric (or gravitational) potential inside the domain due to a charge (or mass) distribution on the boundary B.

1.2.1. The Neumann problem. The heat flux through the boundary is prescribed. It is then to find a harmonic function which is the solution of the equation (1) with prescribed inward normal derivative on B:

$$\frac{\partial u}{\partial n} = f(\theta) \tag{3}$$

As shown later, u is determined up to an additive constant if the solution exists.

It is remarked that, in order to assure the existence of the solution, a necesary condition is

$$\oint_B f(\theta)\, ds = 0 \tag{4}$$

For the present physical problem, this can be intuitively seen that the total flux across the boundary must be zero if a steady state of the heat flow is maintained. A rigorous proof is given later.

Dirichlet and Neumann problems are closely related as shown later (Ex. 1).

1.2 Exterior Problems

For a domain outside a finite number of non-intersecting closed curves, merely to prescribe boundary values on these closed curves are not enough to determine uniquely the harmonic function in this domain. Among other conditions, the behavior of the function at infinity has to be known. For instance, in order to determine the potential of an electric field due to the charges on a number of conductors which are parallel infinite cylinders, we have to consider:

$$\Delta U = 0$$

where

$$U = \alpha_j \quad \text{on} \quad B_j ,$$

where α_j is a constant.

Besides, the harmonic function is assumed to be single-values, and the behavior of the function at infinity has also to be prescribed.

Another example is to find the steady, incompressible flow around a cylinder such as an airfoil of infinite span. The partial differential equationis are:

(A) The Equation of Continuity

$$\frac{\partial u}{\partial x} + \frac{\partial v}{\partial y} = 0 \tag{5}$$

where u, v are (rectangular) velocity components, and

(B) The Equation of Irrotationality*

$$\frac{\partial u}{\partial y} - \frac{\partial v}{\partial x} = 0 \tag{6}$$

The equation (5) assures the existence of a function $\psi(x,y)$ such that

$$u = \frac{\partial \psi}{\partial y}$$

$$v = -\frac{\partial \psi}{\partial x} \tag{7}$$

The function ψ which is defined up to an additive constant has a physical meaning. It represents the mount of the flow across a curve joining a fixed point to the point (x,y), if the density is unity. It is therefore, equal to a constant along a

---

*By the theory of perfect fluid, irrotationality is a consequence of Kelvin theorem, and the fact that the flow at infinity is a uniform. See, for instance Lamb, Hydrodynamics, p. 36.

stream line. On account of this it is called the stream function. From (6), we have

$$\Delta\psi = 0 \tag{9}$$

On the other hand, we can define a function $\varphi(x,y)$ which is called the velocity potential from the equation (6) such that

$$u = \frac{\partial\varphi}{\partial x} \tag{10}$$

$$v = \frac{\partial\varphi}{\partial y} \tag{11}$$

From (5), it follows

$$\Delta\varphi = 0 \tag{12}$$

(C) The boundary condition. The cross section C (the physical body) is a stream line; i.e. ψ = constant, say, zero on C.

(D) At infinity, the flow is uniform

$$\lim_{x^2+y^2\to\infty} (u = \varphi_x = \psi_y) = b$$
$$\lim_{x^2+y^2\to\infty} (v = \varphi_y = -\psi_x) = a \tag{14}$$

One can make the choice of the unit and coordinate axes such that $a = 0$, $b = 1$.

These conditions, as shown later, do not determine the solution uniquely. As u and v are physical realities, it is reasonable to assume that they are single-valued functions of x and y. So is ϕ. The variation of φ along any simple closed curve which encloses only the fluid should zero from the equation (6) by Green's theorem. However, if the curve encloses the body C, the conclusion is no longer valid. But the variation must be the same for any two of such curves, as this is readily seen by equation (6) and by Green's theorem. Such a variation is called the circulation, defined as follows:

(E)

$$\oint_\Gamma udx + vdy = \oint_\Gamma d\varphi = K \tag{15}$$

where Γ is any simple closed curve containing C as its interior. The conditions (13), (14) and (15) now determine the functions φ and ψ (up to a constant) uniquely.

2. SOLUTION OF THE BOUNDARY VALUE PROBLEMS BY FOURIER SERIES

In the case, if C is a unit circle and if the boundary value is expressible in Fourier series, the solution is immediately obtained as follows:

2.1 The Dirichlet Problem

Let

$$x + iy = re^{\theta i}$$

$$\Delta u = 0$$

and

$$u = \sum_{n=0}^{\infty} (a_n \cos n\theta + b_n \sin n\theta) \text{ on C: } r = 1 .$$

Then the solution is

$$(7) \quad u(r,0) = \sum_{n=0}^{\infty} (a_n \cos n\theta = b_n \sin n\theta)\, r^n$$

That (7) converges uniformly and absolutely (and thus $\Delta u = 0$) follows readily from the properties of the power series. That

$$\lim_{r\to 1-0} u(r,\theta) = \sum_{n=0} (a_n \cos n\theta + b_n \sin n\theta) .$$

follows from the Abel's lemma.

2.2 The Neumann Problem

Let

$$\Delta u = 0$$

and

$$u = \sum_{n=0}^{\infty} (a_n \cos n\theta + b_n \sin n\theta) \text{ on C} . \tag{15a}$$

The necessary condition (4) gives that $a_0 = 0$. Thus, the solution of Neumann problem is,

$$u(r,0) = \sum_{n=1}^{\infty} \frac{r^n}{n}(a_n \cos n\theta + b_n \sin n\theta) \tag{16}$$

EXERCISES

(1) If u(x,y) is a harmonic function in a simple connected domain, show that there exists a conjugate function v(x,y) which is defined up to a constant such that

$$\frac{\partial u}{\partial x} = \frac{\partial v}{\partial y} \qquad \frac{\partial u}{\partial y} = -\frac{\partial v}{\partial x}$$

Show that a Dirichlet problem of u induces a Neumann problem of v and vice versa.

(2) Find the solution for the torsion problem of a cylindrical bar with unit circular cross section.

$$\Delta u = 2$$

and u = 0 on the unit circle $C = x^2 + y^2 = 1$.

HINT: Reduce the problem to a Dirichlet problem by setting

$$u = \emptyset + \frac{x^2+y^2}{2}$$

(3) Solution of the Dirichlet Problem for a Circular Ring

Find the single-valued solution of the equation

$$\Delta u = 0 \qquad \text{for} \qquad 1 < r < R$$

where R is the radius of the larger circle such that

$$u = \sum_{n=0}^{\infty} (a_n \cos n\theta = b_n \sin n\theta) \quad , \quad \text{on } r = 1$$

and

$$u = \sum_{n=0}^{\infty} (\alpha_n \cos n\theta = \beta_n \sin n\theta) \quad , \quad \text{on } r = R$$

HINT: Let the solution be of the form:

$$u = A_0 + \frac{(\alpha_0 - A_0)\log r}{\log R} + \sum_{n=1}^{\infty} [(A_n r^n + B_n r^{-n}) \cos n\theta$$

$$+ (C_n r^n + D_n r^{-n}) \sin n\theta]$$

The form of the solution is a consequence of the real part of a logarithmic function plus the real part of an analytic function which possesses a Laurent's expansion. Find A_n, B_n, C_n and D_n in terms of a_n, b_n, α_n, and β_n for $n \geqq 1$.

2.3 Schwarz's and Poisson's Formulas for the Dirichlet Problem in a Unit Circle

Let v be the harmonic conjugate of u where u is the solution of the problem:

$$\Delta u = 0$$

and $u = f(\theta)$, on $r = 1$. $f(\theta)$ is a given continuous function of θ with the period 2π. Then $F(z) = u+iv$ is analytic in the unit circle. Let

$$\zeta = r_0 e^{\varphi i} \quad \text{and } \gamma \text{ be the circle } |\zeta| = r_0$$

We have,

$$F(\zeta) + \bar{F}(\zeta) = 2\, u(r_0,\varphi) .$$

As u is continuous up to the unit circle, it follows:

$$\lim_{r_0 \to 1} u(r_0,\varphi) = f(\varphi) .$$

Thus for any $|z| > r_0$,

$$\frac{1}{2\pi i} \oint_\gamma \frac{F(\zeta)}{\zeta - z}\, d\zeta + \frac{1}{2\pi i} \oint_\gamma \frac{\bar{F}(\zeta)}{\zeta - z}\, d\zeta = \frac{1}{\pi i} \oint_\gamma \frac{u(r_0,\varphi)\, d\zeta}{\zeta - z} \tag{17}$$

The first integral is, by Cauchy's residue formula, $F(z)$. The second integral is evaluated as follows:
Let

$$F(z) = \sum_{n=0}^{\infty} A_n z^n \,, \quad \text{for} \quad |z| < 1.$$

Thus,

$$\frac{1}{2\pi i}\oint_{\gamma} (\overline{F(\zeta)}/\zeta^m)\, d\zeta = \frac{1}{2\pi i} \sum_{n=0}^{\infty} A_n r_0^{\,n-m+1} \int_0^{2\pi} e^{(-n-m+1\varphi i)} d\varphi$$

$$= \begin{cases} \overline{A_0}\,, & \text{for} \quad m = 1, \quad n = 0 \\ 0\,, & \text{for} \quad m > 1 \ \text{and any } n. \end{cases} \qquad (18)$$

Thus

$$\frac{1}{2\pi i}\oint_{\gamma} \frac{\overline{F(\zeta)}}{\zeta - z}\, d\zeta = \frac{1}{2\pi i} \sum_{m=1}^{\infty} \oint_{\gamma} \frac{\overline{F(\zeta)}}{\zeta^m}\, z^{m-1} d\zeta = \overline{A_0}$$

Put

$$F(0) = A_0 = \alpha + \beta i \,.$$

(17) becomes

$$F(z) = -\alpha + \beta i + \frac{1}{\pi i}\oint_{\gamma} \frac{u(r_0,\varphi)}{\zeta - z}\, d\zeta \qquad (18')$$

Let $r_0 \to 1$. We have

$$F(z) = -\alpha + \beta i + \frac{1}{\pi i}\oint_{|\zeta|=1} \frac{f(\varphi)d\zeta}{\zeta - z}$$

Putting

$$z = 0,\ 2\,\alpha = \frac{1}{\pi i}\oint_{|\zeta|=1} \frac{f(\zeta)d\zeta}{\zeta} \,.$$

Thus, (18') yields Schwarz's formula for the analytic function, the real part of which is prescribed on the unit circle:

$$F(z) = \beta i + \frac{1}{2\pi i}\oint_{|\zeta|=1} \frac{\zeta + z}{\zeta(\zeta - z)}\, f(\varphi)\, d\zeta \qquad (19)$$

EXERCISES

(1) Derive Poisson's formula for the solution of the Dirichlet problem of the unit circle from Schwarz's formula (19):

$$u(r,\theta) = \frac{1}{2\pi i}\int_0^{2\pi} \frac{(1-r^2)f(\varphi)d\varphi}{1-2r\cos(\theta-\varphi)+r^2} \qquad (20)$$

HINT: Take the real part of the both sides of (19).

(2) Show that if $f(\varphi)$ is periodic of the period 2π and is continuous, then

$$\lim_{\substack{r\to 1\\ \theta\to\varphi}} [u(r,\theta)] = f(\varphi_o)$$

where $u(r,\theta)$ is given by (20)

HINT: Consider

$$[u(r,\theta) - f(\varphi_o)] = \frac{1}{2\pi}\int_0^{2\pi} \frac{[f(\varphi)-f(\varphi_0)](1-r^2)}{1+r^2-2r\cos(\theta-\varphi)} d\varphi .$$

(3) Boggio's and Dini Formulas

Establish Boggio's formula for the analytic function, the outward normal derivative of the real part of which is prescribed as $f(\varphi)$ on the unit circle such that

$$\int_0^{2\pi} f(\varphi)\, d\varphi = 0:$$

$$F(z) = \alpha + \beta i + \frac{1}{\pi i}\oint_{|\zeta|=1} \frac{f(\varphi)}{\zeta} \log\frac{\zeta}{\zeta - z}\, d\zeta , \quad (\zeta = e^{i\varphi}) . \quad (21)$$

By taking the real part of the both sides, derive Dini's formula for the Neumann problem in the unit circle:

$$u(r,\theta) = \frac{-1}{2\pi}\int_0^{2\pi} f(\varphi)\log[1+r^2-2r\cos(\theta-\varphi)]\, d\varphi \qquad (22)$$

HINT: Consider

$$zF'(z) + \overline{zF'(z)} = 2\left.\frac{\partial u}{\partial r}\right|_{r=1} \quad \text{for} \quad z = e^{\varphi i}$$

2.4 Solution for the Incompressible Flow Around an Infinite Circular Cylinder

Let the cross section of the cylinder be the unit circle $|z| = 1$. Since the complex velocity $W = u+iv$ is a single-valued analytic function of z outside the unit circle and has the normalized limiting value 1 as $z \to \infty$, the function must have the following Laurent's expansion

$$W = 1 + \sum_{n=1}^{\infty} \frac{A_n}{z^n} .$$

Therefore

$$F(z) = \varphi + \psi i = z + A_1 \log z + \sum_{n=2}^{\infty} \frac{A_n}{(1-n)z^{n-1}}$$

From the boundary condition that

$$\psi = 0 \quad \text{on} \quad |z| = 1 \quad \text{or} \quad z = e^{i\theta} ,$$

it follows the identity in θ:

$$\text{Im}\,[e^{i\theta} + A_1 i\theta + \sum_{n=2}^{\infty} \frac{A_n}{1-n} e^{(1-n)\theta i}] = 0 \tag{23}$$

Since ψ is single-valued, $\text{Re}(A_1) = 0$. From (23), $A_2 = -1$ and $A_m = 0$ for $m \geq 3$. Thus one obtains the solution:

$$\emptyset + \psi i = z + \frac{1}{z} + \beta i \log z, \quad \text{where } \beta \text{ is real} \tag{24}$$

or

$$\emptyset(r,\theta) = (r + \frac{1}{r}) \cos\theta - \beta\theta \tag{25}$$

$$\psi(r,\theta) = (r - \frac{1}{r}) \sin\theta + \beta \log r \tag{26}$$

The constant β is determined by prescribing the circulation

$$K = \int_\gamma d\emptyset ,$$

or

$$\beta = \frac{-K}{2\pi} .$$

In other words, there is a family of solutions with an arbitrary assigned circulation K for the circle problem.

3. CONFORMAL MAPPING

In the above discussion, solutions are obtained for the Dirichlet problem and the Neumann problem in a unit circle. These problems for a general domain can be solved by the conformal mapping. An one-to-one analytic mapping is called conformal if it preserves the angle as well as its sense. One advantage to apply this kind of transformation of the independent variables is that the Laplace equation is invariant. That there exists such a mapping which carries a closed Jordan curve into a unit circle with the interior going into the interior was established by Riemann. We shall use this fact to assure the existence of the solution of the general problem. However, for a given simple closed curve, the explicit formula for the mapping function is not available. We shall discuss some special cases which are of paramount importance in many practical problems.

3.1 Möbius Transformations

Consider the mapping

$$W = \frac{az+b}{cz+d} \tag{27}$$

where a, b, and c and d are constants such that $\begin{vmatrix} a,b \\ c,d \end{vmatrix} \neq 0$. Such a transformation defines a conformal mapping between z-plane and W-plane.* It is readily seen that this transformation can always be decomposed into a number of successive mappings of the following types:

(i) $W = Az + B$ where A and B are constants.

(i) $W = \frac{1}{z}$

---

*cf., Section 3.3.1, p. 175.

The transformation of the type (i) is a product of a translation, a rotation and a change of scale. Thus the family of circles and straight lines is invariant under such a transformation. The transformation of type (ii) is an inversion with respect to a unit circle, together with a reflection with respect to the real axis. Thus we establish

Theorem 1. A circle or a straight line is always carried into either a circle or straight line by a Möbius transformation.

3.1.1. A half plane onto a unit circle. The transformation

$$W = \frac{z-a}{z-\bar{a}} \tag{28}$$

where a is a complex number and a its complex conjugate, maps the upper half plane (in z-plane) into the interior of the unit circle in W-plane while the real axis goes into the circumference of the circle provided $Im(a) > 0$.

This is easily established by observing the fact that if z is real, then

$$z-a = \overline{z-a} = z-\bar{a}$$

and that

$$\arg \frac{z-a}{z-\bar{a}} = \arg (z-a) - \arg (z-\bar{a})$$

varies from 0 to 2π as z varies from $-\infty$ to ∞

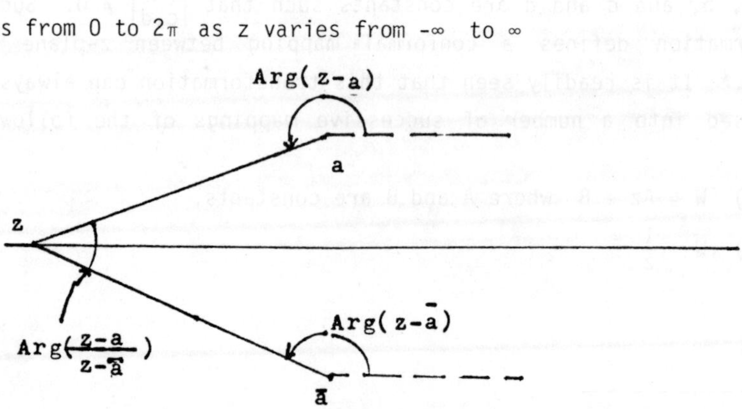

Figure 13

On the other hand, let z be any point on the upper half plane the segment joining a and z would not intersect the real axis. Its image on the W-plane does not intersect the unit circle on W-plane. Since the image of a, which is the origin on the W-plane, lies in the interior of the unit circle in the W-plane, so is the image W of z.

<u>3.1.2. A unit circle onto itself</u>. By similar argument, one can readily show that the transformation

$$W = \frac{z-a}{\bar{a}z-1} e^{\alpha i} ,$$

where

$$|a| < 1 ,$$

and α is real constant, maps a unit circle onto a unit circle. As a matter of fact, it can be shown that these are only transformations which can do this.* As the real and the imaginary parts of a can be chosen arbitrarily as well as α, there are 3-parameter family of transformations which can map a simple closed curve onto a circle.

3.2 The Transformation $W=z^{\alpha}$ where α is a Real Number

It is, in general, necessary, to use many sheets for both z-plane and W-plane in order to define one-to-one mapping. The function $W = z^{\alpha}$ is everywhere analytic except at z = 0 where the angle on the z-plane is magnifying α times in W-plane in the neighborhood or W = 0. For instance, a sector with the vertex at z=0 of an angle φ is mapped conformally onto a sector with the vertex W = 0 of an angle $\alpha\varphi$ except at the vertex. This result is very useful in the later discussions, such as, the conformal mapping of a polygon onto a unit circle.

---

*For Schwarz Lemma, See Caratheodory, Conformal Mapping, Cambridge Tracts, No. 28, p. 39, 1932.

EXERCISES

(1) Show that $W = z^2$ maps the first quadrant onto the upper half plane. Thus, find a mapping function which maps the unit circle onto the first quadrant.

(2) Find a mapping function which maps a half circle onto a unit circle.

HINT: Let the diameter intersect the circle at A and B. Find a Möbius transformation which carries B into the infinity while A is remained unchanged. At A, the mapping is conformal. So the image of the half circle is a quarter of the plane. Then apply the result of Ex. 1.

(3) By a similar procedure as in Ex. 2, find a mapping function which carries a region bounded by two circular arcs onto a unit circle.

(4) Discuss the following transformations:

(a) $W = e^z$ (B) $W = \sin z$.

(5) Show that $W = z + \frac{1}{z}$ maps the domain outside the unit circle in z-plane onto the domain outside a slit $-2 \leqq W \leqq 2$ in W-plane with the point at infinity invariant.

(6) By Ex. 5, find the incompressible flow around a slit of length 4 with an angle of "attack" to the negative real axis.

HINT: Show that if the flow has an angle of attack α i.e., a uniform flow at infinity with velocity components $u = \cos\alpha$, $v = \sin\alpha$, the solution of the circle problem in z-plane is

$$\emptyset + i\psi = ze^{-\alpha i} + \frac{i}{ze^{-\alpha i}} - \frac{Ki}{2\pi}\log z \qquad (29)$$

Now consider the transformation:

$$= e^{-\alpha i}\left(z + \frac{1}{z}\right) \qquad (30)$$

which maps the unit circle into the slit in the problem. Then the uniform flow with angle of attack α in z-plane is transformed into one parallel to the real axis in z-plane. Thus, the solution of the present problem is reduced to solve z in terms of ζ. Make the choice of K such that the velocity

$$\frac{d}{d\zeta}(\emptyset + i\psi) = \frac{d(\emptyset+i\psi)}{dz} / \frac{d\zeta}{dz}$$

is finite at the trailing edge of the slit. This condition is due to Kutta-Joukowsky. It is then seen that the velocity at the leading edge of the slit is infinite.

(7) Joukowsky Airfoils

Show that the function

$$\zeta = z_1 + \frac{a^2}{z_1} \qquad (0 < a < 1) \qquad (31)$$

maps the exterior of the circle $|z-\delta i|^2 = a^2+\delta^2$ (where δ is a real parameter) onto the exterior of the arc of the circle in ζ-plane as indicated in the following Figure 14.

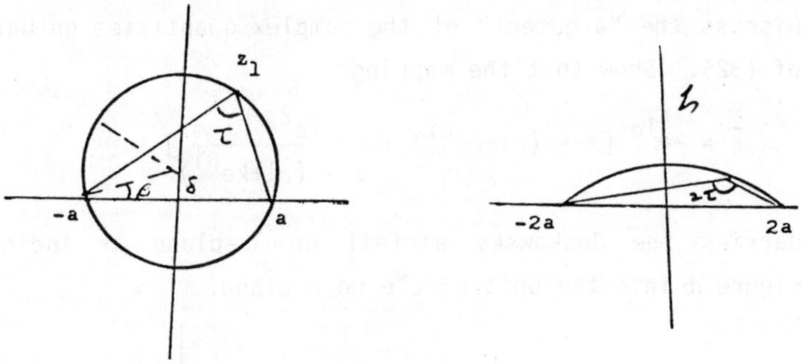

Figure 14

The circle

$$|z - (\delta i + ke^{\beta i})| = 1 ,$$

where

$$\beta = \tan^{-1}\frac{\delta}{a} \quad \text{and} \quad k = 1 - \sqrt{a^2+\delta^2} ,$$

is thus mapped into a Joukowsky airfoil.

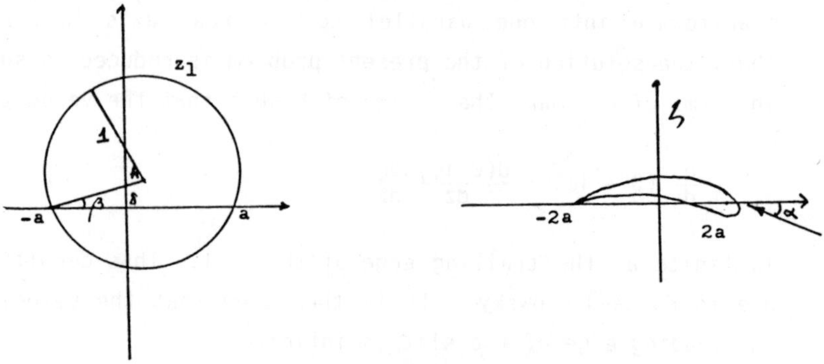

Figure 15

Find the flow around a Joukowsky airfoil with an angle of attack α as indicated in Figure 15.

<u>HINT</u>: The mapping (31) can be rewritten as

$$\frac{\zeta+2a}{\zeta-2a} = \left(\frac{z+a}{z-a}\right)^2 \tag{32}$$

(8) Discuss the "argument" of the complex quantities on both sides of (32). Show that the mapping

$$\bar{\zeta} = -e^{i\alpha}\left[z - (\delta i + ke^{\beta i}) + \frac{a^2}{z - (\delta i + ke^{\beta i})}\right]$$

carries the Joukowsky airfoil on ζ-plane as indicated in Figure 16 into the unit circle on z-plane.

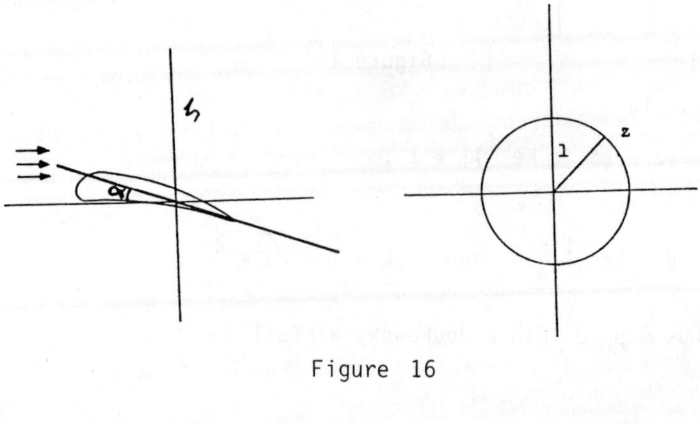

Figure 16

Find the flow around the unit circle on z-plane so that the corresponding flow is a uniform flow at infinity parallel to the real axis on ζ-plane. Determine the circulation so that the velocity of the flow is finite at the trailing edge of the airfoil in ζ-plane.

3.3 The Schwarz - Christoffel's Formula

3.3.1. Theorem 1. Let $w = f(z)$ be analytic in $|z - a| \leq \varepsilon$, and $f'(a) \neq 0$. The mapping is conformal in the neighborhood of $z = a$.

Proof. Let $w = \xi + \eta i$ and $z = x + yi$. The Jacobian at $z = a$ is

$$J = \begin{vmatrix} \frac{\partial \xi}{\partial x}, & \frac{\partial \eta}{\partial x} \\ \frac{\partial \xi}{\partial y}, & \frac{\partial \eta}{\partial y} \end{vmatrix} = |f'(a)|^2$$

By Cauchy-Riemann differential equations. By the implicit functions theorem, the function defines an one-to-one mapping between the neighborhood of $z = a$ and that of $w = f(a)$.
If dz_1 and dz_2 are two arcs at $z = a$ in z-plane, the angle between dz_1 and dz_2 is

$$[\arg(dz_2) - \arg(dz_1)]$$
$$= \arg(f'(a)dz_2) - \arg(f'(a)dz_1)$$
$$= \arg(dw_2) - \arg(dw_1)$$

where dw_2 and dw_1 are the corresponding arcs in w-plane.

3.2.2 Theorem 2. (The converse of Theorem 1) Let $w = f(z)$ satisfy the condition of analyticity as in the Theorem 1. If $f'(a) = 0$, then the mapping is one-to-many from z-plane to w-plane. Let us first establish the following

Lemma. Let $g(z)$ be analytic in a domain D bounded by a simple closed curve γ. Let $g(z)$ be continuous and $= 0$ on γ.

$$\frac{1}{2\pi i}\oint_\gamma \frac{g'(z)}{g(z)}\,dz = \text{the number of zeroes of } g(z) \text{ inside } \gamma.$$

(A zero of multiplicity m is counted as m times)

Let g(z) have a zero at z = b of multiplicity m. g'(z) has a zero of multiplicity m-1 at z = b. Thus g'(z)/g(z) has a simple pole at z = b and has the residue at z = b:

$$\lim_{z\to b} \frac{g'(z)(z-b)}{g(z)} = m\,,$$

since

$$g(z) = A_m(z-b)^m + A_{m+1}(z-b)^{m+1} + \ldots.$$

$$(z-b)g'(z) = mA_m(z-b)^m + (m+1)A_{m+1}(z-b)^{m+1} + \ldots.$$

As the integral, by Cauchy's residue theorem, is equal the sum of the residues at the zeroes of g(z), the lemma is immediately established.

Proof to the Theorem 2. Suppose $f(z) \neq$ constant. (Otherwise the theorem is trivial). Let ε be small enough such that $f'(z) \neq 0$ for $0 < |z-a| \leq \varepsilon$ and on γ: $|z-a| = \varepsilon$, $|f(z)-f(a)| > 0$. Therefore there exists a value η such that $|f(z)-w_1| > 0$ for $|z-a| = \varepsilon$ and $|f(a)-w_1| < \eta$.

Let m_1 be the number of zeroes of $f(z)-w_1$ inside γ. ($w_1 \neq f(a)$). On the account of $f'(z) \neq 0$, all their roots, if any, are simple. Let m be the multiplicity of the zero of $[f(z)-f(a)]$ at z = a. Now, as $m \geqq 2$, it suffices to show that $m_1 = m$ provided that η is small enough.

According to the lemma,

$$m-m_1 = \frac{1}{2\pi i}\oint_\gamma f'(z)\left\{\frac{1}{f(z)-f(a)} - \frac{1}{f(z)-w_1}\right\}dz$$

$$= \frac{1}{2\pi i}\oint \frac{f'(z)\{f(a)-w_1\}}{\{f(z)-f(a)\}\{f(z)-w_1\}}dz \qquad (33)$$

Put

$$A = \min_{z \text{ on } \gamma} |f(z) - f(a)| > 0 \quad .$$

Choose η such that $\eta < \frac{A}{2}$. Then for z on γ.

$$|f(z) - w_1| > |f(z) - f(a)| - |f(a) - w_1|$$

Put, again,

$$\max_{z \text{ on } \gamma} |f'(z)| = B \ .$$

It is clear from (33) that $|m-m_1| \leq \frac{2B\varepsilon}{A^2} \eta$. Take η smaller yet, if necessary; namely,

$$\eta < \frac{A^2}{2B\varepsilon}$$

One concludes from the fact that $|m-m_1| < 1$, the two integers m and m_1 must be equal.

EXERCISE.

(1) Prove the fundamental theorem of algebra that any polynomial with complex coefficients has at least one complex root.

HINT: Let the polynomial be P(z). Consider

$$\frac{1}{2\pi i} \oint_{|z|=R} \frac{P'(z)}{P(z)} dz$$

Show that the integral is independent R if the latter is large enough. Take the limit of the integral as $R \to \infty$.

3.3.3. Schwarz-Christoffel's Formula for Mapping the Upper Half Plane onto a Polygon. Consider

$$w = A \int_0^z \frac{d\zeta}{(\zeta-a_1)^{\beta_1}(\zeta-a_2)^{\beta_2} \ldots\ldots(\zeta-a_n)^{\beta_n}} + B \ , \qquad (34)$$

where

$$\beta_1 + \beta_2 + \ldots + \beta_n = 2 , \quad (-1 < \beta_j < +1) ,$$

and $a_1, a_2 \ldots a_n$ are real.

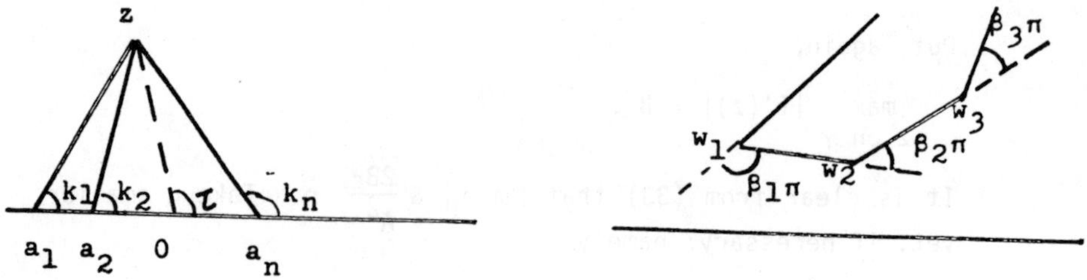

Figure 17

Put

$$z - a_j = \rho_j e^{ik_j} , \quad 0 \leqq k_j \leqq \pi ,$$

Define

$$(z-a_j)^{\beta_j} = \rho_j^{\beta_j} e^{i\beta_j k_j}$$

The integrand is then a single-valued analytic function of z and is finite ($\neq 0$) in the upper half plane except at a_j. So is w as a function of z. Besides, w is also continuous up to the whole real axis in the z-plane. Let $w_j = w(a_j)$. For $a_j < z < a_{j+1}$ on the real axis, arg $(\frac{dw}{dz})$ is a constant. It is then clear from the proof of Theorem 1 that the segment on z-plane is mapped into a segment on w-plane between w_j and w_{j+1}. At the point a_j, the mapping locally behaves like

$$w = \frac{1}{1-\beta_j} (z-a_j)^{1-\beta_j} + \ldots$$

As z moves on the real axis crossing a_j, w turns an angle $\beta_j\pi$ as indicated in the Fig. 5. Since $\beta_1 + \beta_2 + \ldots + \beta_n = 2$, the real axis in z-plane is mapped into a polygon. The numbers β_j are restricted to the range $-1 < \beta_j < +1$.

Again, by Theorem 1, the mapping is conformal* at any point other than a_j on the real axis in z-plane. The points on the upper half of the z-plane are mapped into points interior of the polygon as indicated in Fig. 7. That the images of all points in the upper half z-plane fill up the interior of the polygon follows from the continuity. Otherwise there would be a boundary point inside the polygon corresponding a point in the upper half z-plane.

EXERCISE.

(1) Find the mapping function which maps the upper half z-plane onto a rectangle with the points $z = \pm 1, \pm \frac{1}{k}$ into the corners $(k < 1)$.

ANSWER.

$$w = \int_0^z \frac{dz}{\sqrt{(1-z^2)(1-k^2z^2)}}$$

(2) Degenerate polygons.

(a) The infinite strip. Consider the mapping as indicated in Fig. 18.

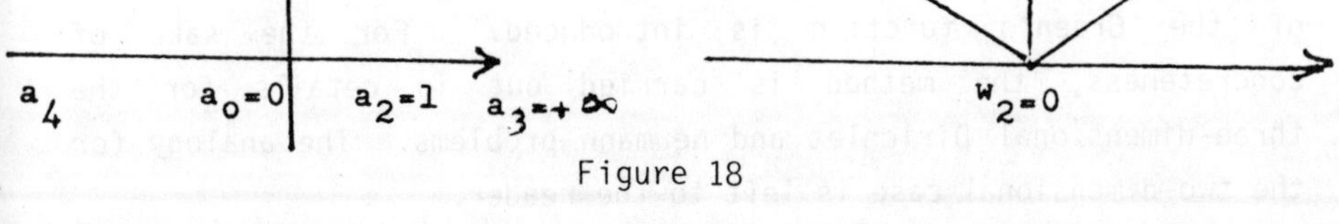

Figure 18

*That the mapping is locally conformal is immediately seen. That the mapping is one to one in the large is intuitively seen from the present geometrical consideration. For a rigorous derivation of Christoffel-Schwarz's formula, see, Copson, Theory of Functions of a Complex Variable, pp. 193-198.

Let w_1 and w_3 tend to infinity. Find the mapping function.

ANSWER.

$$w = \log z$$

(b) A half line in an infinite strip.

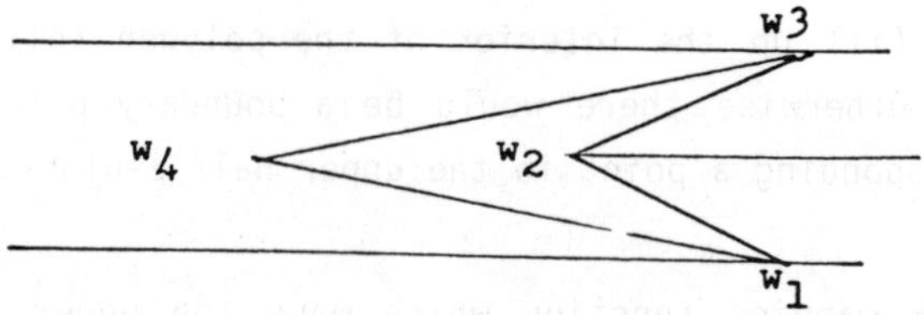

Figure 19

Map the domain with vertices w_1, w_2, w_3, and w_4 into the upper half plane such that $a_1 = -1$, $a_3 = 1$, $a_4 = \infty$. Let w_1, w_3 and w_4 tend to infinity. Find the mapping function.

ANSWER.

$$w = -\frac{1}{2}\log(z^2-1) + \pi i \quad .$$

4. GREEN'S FUNCTIONS AND METHOD OF IMAGES

The preceding method is successful only for attacking the two-dimensional boundary value problems. In order to deal with the problems in higher dimensions and other types of partial differential equations (not limited to the second order) the notion of the Green's function is introduced. For the sake of concreteness, the method is carried out in details for the three-dimensional Dirichlet and Neumann problems. The analogy for the two-dimensional case is left to the reader.

Consider the boundary value problem in a bounded domain D:

$$\Delta u = 0 \; , \quad \Delta = \frac{\partial^2}{\partial x^2} + \frac{\partial^2}{\partial y^2} + \frac{\partial^2}{\partial z^2}$$

with L(u) prescribed on the boundary B of D, where L(u) = u or $\frac{\partial u}{\partial n}$.

The Green's function of the domain D with respect to a fixed point $Q(\xi, \eta, \zeta)$ in D can be defined physically as the potential due to a unit charge while the surface B, as a conductor is grounded. Symbolically, the Green's function is defined as $u = G(x,y,z; \xi,\eta,\zeta)$ such that

(i) $\Delta u = \delta(x-\xi, \; y-\eta, \; z-\zeta)$ where δ is the three-dimensional δ -function.

(ii) $L(u) = 0$ on B. ($L(u)$ = A constant for the Neumann problem)

The rigorous definition is given as follows:

<u>Definition</u>: A function $G(x,y,z; \xi,\eta,\zeta)$ is defined as the Green's function of D with respect to the point $Q(\xi,\eta,\zeta)$ if the following condition is satisfied.

(1) $\Delta G = 0$ except $(x,y,z) = (\xi,\eta,\zeta)$ (35)

(2) $L(G) = 0$ on $B(L(G)$ = a constant for the Neumann problem) (36)

(3) G has a singularity of the type

$$\frac{1}{4\pi r} \text{ where } r = \{(x-\xi)^2 + (y-\eta)^2 + (z-\zeta)^2\}^{\frac{1}{2}} \tag{37}$$

at Q; namely

$$G = \frac{1}{4\pi r} - g(x,y,z; \xi,\eta,\zeta) \tag{38}$$

where g is a harmonic function in D for fixed (ξ,η,ζ) such that

$$L[g(x,y,z; \xi,\eta,\zeta)] = L\left(\frac{1}{4\pi r}\right)$$

The existence of the Green's function or the function g is beyond the scope of the present treatment. The reader is referred to the standard text; for instance, see Kellogg, Potential Theory.

<u>Example 1.</u> <u>The Green Function for the Half Space Bounded by a Plane</u>

Consider the Dirichlet Problem. Let the domain be $z \geqq 0$. B is then the xy plane. At the point $Q(\xi,\eta,\zeta)$

where $\zeta > 0$, a positive unit charge is placed. In order to keep the potential zero at $z = 0$, a negative charge is placed at $Q'(\xi,\eta,-\zeta)$. Thus the Green's function is

$$G(x,y,z;\ \xi,\eta,\zeta) = \frac{1}{4\pi}\left[\frac{1}{d(P,Q)} - \frac{1}{d(P,Q)}\right]$$

where

$$d(P,Q) = \{(x-\xi)^2 + (y-\eta)^2 + (z-\zeta)^2\}^{\frac{1}{2}}$$

$$d(P,Q') = \{(x-\xi)^2 + (y-\eta)^2 + (z+\zeta)^2\}^{\frac{1}{2}}$$

This is called the <u>method of images</u>.

<u>Example 2.</u> <u>The Green's Function of a Unit Sphere</u>.

First of all, the Laplace equation has a certain semi-invariant property under the transformation of coordinates by an inversion with respect to the unit sphere. According to the exercise 1, Ch. I the Laplace equation in spherical coordinates can be written as

$$L_{r,\theta,\varphi}(u) = \left[\frac{1}{r}\frac{\partial^2}{\partial r^2} + \frac{1}{r^3}\frac{1}{\sin^2}\frac{\partial^2}{\partial \theta^2} + \frac{1}{r^3 \sin\varphi}\frac{\partial}{\partial \varphi}\left(\sin\ \frac{\partial}{\partial \varphi}\right)\right](ru) = 0 \tag{39}$$

Set

$$r = \frac{1}{r_1} \tag{40}$$

$$v = ru \tag{41}$$

It is immediately verified that

$$L_{r,\theta,\varphi}(u) = r_1^5\, L_{r_1,\theta,\varphi}(v) \tag{42}$$

In other words, with a specific change of the dependent variables (41), the Laplace equation is unchanged under the transformation of coordinates induced by the inversion with respect to the unit sphere. As a plane is mapped to a sphere by an inversion and vice versa, a boundary value problem on a domain bounded by a sphere is thus reduced to one on the domain bounded by a plane. This fact is, in particular, very useful for the Green's function. However, it ought to be remarked that (41) does not, in general, give directly the Green's function of a sphere from that of the half space. A proper factor which is a function of the fixed point Q must be introduced in order that the singularity at Q is a correct one.

Let the half space be $z \geqq \frac{1}{2}$ be mapped onto the domain bounded by a unit sphere with the center at (0, 0, 1) by an inversion (with respect to unit sphere $x^2+y^2+z^2 = 1$). The corresponding points and boundaries on a plane section are indicated in the following Fig. 20. Since the Green's function for the unit sphere is evidently sysmetrical with the line CQ as its axis we consider only the points P_1 on OCQ plane.

The plane section

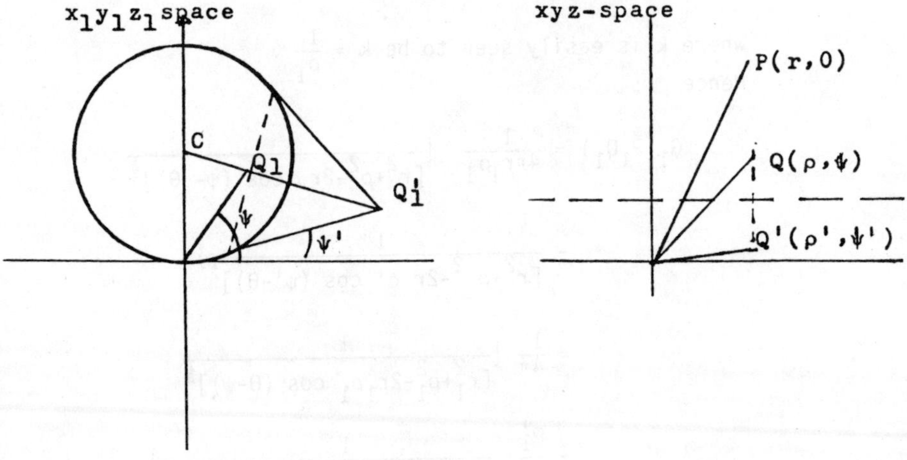

Figure 20

On this plane section, let the polar coordinates (r,θ) and (ρ,ψ) be used respectively for the point P and Q.

The corresponding points on $x_1 y_1 z_1$ - space are denoted by the subscript "1". Accordingly, the transformation induced by inversion is

$$r = \frac{1}{r_1}$$

$$\rho = \frac{1}{\rho_1} \tag{43}$$

$$\rho' = \frac{1}{\rho'_1}$$

It can be readily shown that the points Q_1 and Q_1' form the harmonic conjugates with respect to the circle $x_1+y_1+z_1 = 2z_1$; namely,

$$CQ \cdot CQ_1' = 1 .$$

The Green's function of this unit sphere is

$$G_1(P_1,Q_1) = \frac{k}{r_1} \; G(P,Q) ,$$

where k is easily seen to be $k = \dfrac{1}{\rho_1}$.
Hence

$$G_1(P_1Q_1) = \frac{1}{4\pi r_1\rho_1} \Big\{\frac{1}{[r^2+\rho^2-2r\,\rho\cos(\psi-\theta)]^{\frac{1}{2}}}$$

$$-\frac{1}{[r^2+\rho'^2-2r\,\rho'\cos(\psi'-\theta)]^{\frac{1}{2}}}\Big\}$$

$$= \frac{1}{4\pi}\Big\{\frac{1}{[r_1^2+\rho_1^2-2r_1\rho_1\cos(\theta-\psi)]^{\frac{1}{2}}}$$

$$-\frac{\rho_1'}{\rho_1}\,\frac{1}{[r_1^2+\rho_1'^2-2r_1\rho_1'\cos(\theta-\psi')]^{\frac{1}{2}}}\Big\}$$

By (43),

$$G_1(P_1,Q_1) = \frac{1}{4\pi}\Big[\frac{1}{d(P_1Q_1)} - \frac{\rho_1'}{\rho_1}\,\frac{1}{d(P_1,Q_1')} \tag{44}$$

One can show that $\frac{\rho_1'}{\rho_1} = \frac{\rho_1}{d(CQ_1)}$. This can either be seen by the fact that the Green's function vanishes on the boundary and that $CQ \cdot CQ_1' = 1$. Or, by using the rectangular coordinates (a,b) for Q_1 and (a',b') for Q_1', it follows that $a' = ka$, $b' = 1+k(b-1)$, where

$$k = \frac{1}{a^2+(b-1)^2}$$

and

$$\frac{\rho_1'^2}{\rho_1^2} = \frac{\left(\frac{a}{a^2+(b-1)^2}\right)^2 + \left(\frac{b-1}{a^2+(1-b)^2}\right)^2}{a^2+b^2} = \frac{1}{a^2+(b-1)^2}$$

$$= \frac{1}{d^2(C,Q_1)}$$

The interpretation of the Green's function of a unit sphere in terms of the geometrical quantities is now evident. In the language of physics, this means that a unit charge is put at the point Q_1 and a negative charge of intensity $\frac{R}{d(C,Q_1)}$ is placed on the harmonic conjugate point Q_1' where R is the radius of the sphere.

EXERCISE.

(1) Construct the Green's fucntion for the Dirichlet problem on a unit circle.

ANSWER.

$$\frac{-1}{4\pi} \log \frac{1+\rho^2 r^2 - 2r\rho \cos(\theta-\psi)}{r^2+\rho^2-2r\rho\cos(\theta-\psi)}$$

when (r,θ) and (ρ,ψ) are respectively the polar coordinates of P and Q.

(2) Find the three-dimensional Green's function of the domain $0 \leq z \leq 1$ by the method of images.

ANSWER.

$$\sum_{k=-\infty}^{\infty} \frac{1}{4\pi} \left[\frac{1}{\{(x-\xi)^2+(y-\eta)^2+(z-\zeta+2k)^2\}^{\frac{1}{2}}} - \frac{1}{\{(x-\xi)^2+(y-\eta)^2+(z+\zeta+2k)^2\}^{\frac{1}{2}}}\right]$$

4.1 Green's Function for the Neumann Problem of a Unit Sphere

It is remarked that the above method of inversion fails in obtaining the Green's function of the Neumann problem. Because if u is the Green's function for the half space, by inversion the function v as given by (41) does not satisfy the boundary condition of the Neumann problem. However, this problem can be solved by means of spherical harmonics P_n $(\cos\varphi)$ as introduced in Chapter II pp. 47-50.

Let (r,φ) be the polar coordinates of P on the plane section OPQ by taking the center O as the origin. Let the polar coordinates of Q be $(\rho,0)$. The Laplace equation is reduced to (by means of the rotational symmetry):

$$r^2 \frac{\partial^2 u}{\partial r^2} + 2r \frac{\partial u}{\partial r} + \frac{1}{\sin\varphi} \frac{\partial}{\partial\varphi}\left(\sin\varphi \frac{\partial u}{\partial\varphi}\right) = 0$$

Put

$$u = \frac{1}{d(P,Q)} + \frac{1}{\rho} \frac{1}{d(P,Q')} + v$$

$$= \frac{1}{\{r^2+\rho^2-2r\rho\cos\varphi\}^{\frac{1}{2}}} + \frac{1}{\{\rho^2 r^2+1-2r\rho\cos\varphi\}^{\frac{1}{2}}} + v \qquad (47)$$

Since the first two terms in the right hand are harmonic functions with the proper singularity at P = Q, v satisfies the same equation (46) as well as u. One must take the boundary condition

$$\frac{\partial u}{\partial n} = \left.\frac{\partial u}{\partial r}\right|_{r=1} = \text{constant}$$

such the <u>necessary condition for the Neumann problem</u> is satisfied:

$$\iint_S \frac{\partial v}{\partial n}\, ds = 0 \qquad (48)$$

where S is the surface of the domain and n is the outward normal at the point on the surface. In the present case, it suffices, as one sees later, to take the constant as -1. Thus

$$\left.\frac{\partial v}{\partial r}\right|_{r=1} = \frac{1}{\{1+\rho^2-2\rho\cos\varphi\}^{\frac{1}{2}}} - 1 \tag{49}$$

According to Ex. 5, p. 50, the boundary condition can be expressed in terms of the spherical harmonics $P_n(\cos\varphi)$:

$$\left.\frac{\partial v}{\partial r}\right|_{r=1} = \sum_{n=1}^{\infty} \rho^n P_n(\cos\varphi) \tag{50}$$

Therefore the solution is readily seen to be

$$v(P,Q) = \sum_{n=1}^{\infty} \frac{1}{n} r^n \rho^n P_n(\cos\varphi)$$

$$= \int_0^{\infty} \left[\frac{1}{\{r^2\rho^2+1-2r\rho\cos\varphi\}^{\frac{1}{2}}} - 1\right] \frac{dr}{r}$$

$$= \log\left[\frac{2}{1-r\rho\cos\varphi + \{r^2\rho^2+1-2r\rho\cos\varphi\}^{\frac{1}{2}}}\right] \tag{51}$$

Hence the Green's function for the Neumann problem of a sphere with the radius a is

$$4\pi H(P,Q) = \frac{1}{d(P,Q)} + \frac{a}{\rho}\frac{1}{d(P,Q')} + \frac{1}{a}\log\frac{2a^2}{a^2 - r\rho\cos\varphi + \rho d(P,Q')} \tag{52}$$

Where

$\rho = d(O,Q)$, O = the center of the sphere

and

$\varphi = \angle QOP$

5. SOLUTIONS OF THE BOUNDARY VALUE PROBLEMS BY MEANS OF THE GREEN'S FUNCTION

In the following discussion, the domain is always assumed to be regular. In order to give a precise definition of such a domain,

regular surface element.

Definition 1. If, for some orientation of the coordinate axes, a surface element possesses a standard representation

$$z = z(x,y)$$

where the points lies in a closed region R in xy-plane, bounded by a closed Jordan curve such that z is continuously differentiable* in R, then it is called a regular surface element.

The boundary of a regular surface element consists of a finite regular arcs, called the edges and the intersection of the edges is called the vertex.

Definition 2. A regular surface consists of a finite number the regular surface elements such that

(i) Two elements have in common a vertex or an edge but no other points;

(ii) Three or more elements have, at most, a vertex in common;

(iii) Any two of the elements are the first and last of a chain, such that each has an edge in common with the next;

(iv) All the elements having a vertex in common for a chain such that each has edge containing the vertex in common with the next.

Definition 3. A regular surface is said to be closed if all the edges of the regular surface elements belong to two of the elements.

---

*z is said to be continuously differentiable in R if (i) it is continuous in the closed region R and (ii) both $\frac{\partial z}{\partial x}$ and $\frac{\partial z}{\partial y}$ exist at all interior points of $\bar{R}$ and coincide with a function which is continuous in the closed region $\bar{R}$.

Definition 4. A regular domain of space is defined to be a bounded closed domain whose boundary is a closed regular surface.

5.1 The Green-Gauss Divergence Theorem

Let D be any regular domain bounded by S. Let (X,Y,Z) = F be continuous and piecewise continuously differentiable in D. Then

$$\iiint_D \left(\frac{\partial X}{\partial x} + \frac{\partial Y}{\partial y} + \frac{\partial Z}{\partial z}\right) dV = \iint_S (X \cos \alpha + Y \cos \beta + Z \cos \gamma) dS \quad (53)$$

where (cos α, cos β, cos γ) is the direction cosine of the outward normal. The theorem is sometimes put in the following vector form:

$$\iiint_D \operatorname{div}(\vec{F})\, dV = \iint_S (\vec{F} \cdot d\vec{S}) \quad (3')$$

The proof for simple domains are given in most of the texts of advanced calculus. For the proof in the regular domain, the reader is referred to Kellog, Foundations of Potential Theory, pp. 113-119.

5.1.1. Green's formulas and their consequences. Set

$$X = fg_x, \quad Y = fg_y, \quad Z = fg_z$$

The divergence theorem immediately gives Green's First Formula.

$$\iiint_D f\Delta g\, dV + \iiint_D (f_x g_x + f_y g_y + f_z g_z)\, dV = \iint_S f \frac{\partial g}{\partial n}\, dS \quad (54)$$

By interchanging f and g and taking the difference of the two expressions, it follows immediately:

Green's Second Formula.

$$\iiint_D (f\Delta g - g\Delta f) dV = \iint_S \left(f \frac{\partial g}{\partial n} - g \frac{\partial f}{\partial n}\right) dS \quad (55)$$

Let $Q(\xi,\eta,\zeta)$ be any fixed interior point of D. Set

$$r_{PQ}^2 = (x-\xi)^2 + (y-\eta)^2 + (z-\zeta)^2 .$$

Let δ_ε be the open sphere $r_{PQ} < \varepsilon$, (which is inside D). Denote

$$D' = D - \delta_\varepsilon$$

The limiting case of Green's second formula for the domain D' as $\varepsilon \to 0$ (when $g = \frac{1}{r_{PQ}}$) gives

Green's Third Formula.

$$-4\pi f(P) = \iiint_D \frac{1}{r_{PQ}} \Delta f(Q)\, dV_Q + \iint_S [f \frac{\partial}{\partial n}(\frac{1}{r_{PQ}}) - \frac{1}{r_{PQ}} \frac{\partial f}{\partial n}]\, dS \quad (56)$$

The following exercises are important results in the theory of harmonic functions.* Most of them follow as the consequences of Green's formulas. Generalizations to the general elliptic equations are made (concerning the uniqueness theorem). The reader should have no difficulty to work out in details.

EXERCISES.

(1) The necessary condition for the Neumann problem.

Show by (54), that if g is harmonic in D, and if g is continuously differentiably in D, then

$$\iint_S \frac{\partial g}{\partial n}\, dS = 0 \quad (57)$$

HINT: Let f = 1.

(2) The Gauss' theorem of the mean for harmonic functions.

Show, by (56), that if f is harmonic in D, the arithmetic mean of f over any sphere in D is equal to the value of f at the center of the sphere; namely

$$f(\xi,\eta,\zeta) = \frac{1}{4\pi r^2} \iint_S f(x,y,z)\, dS \quad (58)$$

---

*A function is said to be harmonic in D, if

(i) it admits continuous second partial derivatives in the interior of the domain D (simply to say in D).

(ii) it satisfies the Laplace equation g = 0 in D; and

(iii) it is continuous in the closed domain $\bar{D}$.

where S is the sphere,

$$(x-\xi)^2 + (y-\eta)^2 + (z-\zeta)^2 = r^2 ,$$

and r is a constant.

(3) The extremum values of a harmonic function.

Show that if D is any closed connected* domain (regular or not) and if f is harmonic in D and is not a constant, then f attains its maximum and minimum values on the boundary. Consequently, the solution of the Dirichlet problem, if any, is unique.

HINT: Let M be the maximum value of f. Let E be the set of all points P such that f(P) = M. Show, by Gauss' theorem of the mean, that E can not contain any interior point of D. (cf. J.L. Walsh's proof on p. 223 in Kellog's Potential Theory).

(4) The uniqueness of the Neumann problem in a regular domain.

Show by (54), that the solution of the Neumann problem in a regular domain, if any, is unique (up to an additional constant).

HINT: Consider the harmonic function f, and g = f

(5) The extremum values and the uniqueness theorem of general elliptic partial differential equation.

Show that the solution of the partial differential equation

$$\frac{\partial^2 u}{\partial x^2} + \frac{\partial^2 u}{\partial y^2} + a(x,y) \frac{\partial u}{\partial x} + b(x,y) \frac{\partial u}{\partial y} + c(x,y) \quad u = 0 \quad (59)$$

where a,b,c are continuous a closed connected domain and $c < 0$, has neither a positive maximum value, nor a negative minimum value in the interior. The solution of the Dirichlet problem, if exists, is therefore unique.

---

*A domain is said to be connected in space if for any two points in the interior of the domain, one can always join them by a Jordan curve which lies completely in the interior.

HINT: Show that at an interior point (ξ, η) where the maximum value of u occurs.

$$\frac{\partial u}{\partial \xi} = \frac{\partial u}{\partial \eta} = 0 \quad \text{and} \quad u_{\xi\xi} + u_{\eta\eta} \leqq 0 \; .$$

Then the equation (59) is satisfied only if $u \leqq 0$.

(6) Show that the results in Exercise 5 are valid under the following conditions:

(i) $c \leqq 0$

(ii) The equation (59) is replaced by:

$$Au_{xx} + 2\,Bu_{xy} + Cu_{yy} + au_x + by_y + cu = 0 \qquad (60)$$

Where A, B, C and continuous in $\bar{D}$ and

$$B^2 - AC < 0 \,, \quad \text{and} \quad A > 0 \; .$$

HINT: For (i) let

$$u = u_o(x,y)\; v(x,y)$$

where

$$u_o(x,y) = \alpha - e^{\beta x} \quad ,$$

α and β are constants to be chosen such that

$$u_o(x,y) > 0 \; .$$

and the coefficient of v in the transformed equation is < 0 in $\bar{D}$. For the generalization (ii), show that

$$Au_{xx}\,\xi^2 + 2Bu_{xy}\,\xi\eta + Cu_{yy}\,\eta^2$$

is a definite form at an interior point which gives the extremum value of u.

5.2 The Solutions of Dirichlet and Neumann Problems

Green's second formula in the above sectioni yields immediately the following:

Theorem. If u is harmonic (and continuously differentiable) in D bounded by S and if the Green's function G(Q,P) exists and is continuously differentiable in D except at a neighborhood of the point P, then

$$u(P) = -\iint_S u(Q) \frac{\partial}{\partial n_Q} G(Q,P)\, dS_Q \tag{61}$$

It is remarked that the condition on the first partial derivatives of u can be removed by a limit process. On the other hand, for any prescribed continuous function u(Q) on S, the expression in the right-hand side of (61) can be shown, under a limitation of D, to represent a harmonic function and to have the limit

$$\lim_{P \to Q} u(P) = u(Q) \tag{62}$$

where P belongs to D. An investigation of these problems is beyond the scope of the present treatments. However, such results can be established immediately for special domains such as a sphere.

Proof of the Theorem. Let D_1 be a small sphere bounded by S_1 with the center at P. Applying the Green's second formula to the domain $D' = D - D_1$ for the functions u and G, one has

$$\iint_S \left(u \frac{\partial G}{\partial n} - G \frac{\partial u}{\partial n}\right) dS = -\iint_{S_1} \left(u \frac{\partial G}{\partial n} - G \frac{\partial u}{\partial n}\right) dS \tag{63}$$

As S_1 shrinks to the point P,

$$\iint_{S_1} G \frac{\partial u}{\partial n}\, dS \to 0 .$$

Consider

$$-\iint_{S_1} u \frac{\partial G}{\partial n}\, dS = \iint_{S_1} u\left(\frac{\partial}{\partial r} \frac{1}{4\pi r} - \frac{\partial g}{\partial r}\right) dS$$

where

$$r = d(P,Q)$$

and

$$G = \frac{1}{4\pi r} - g(Q,P) \quad .$$

On S_1,

$$u(Q) = u(P) + o(r) \; .$$

Thus

$$-\iint_{S_1} u(Q)\, \frac{\partial G}{\partial n}\, dS \to -u(P) \quad \text{as} \quad r \to 0.$$

Now, on S, G = 0. Let $r \to 0$, one obtains (61) immediately from (63).

By the same reasoning, the corresponding formula for <u>Neumann problem is</u>

$$u(P) = \iint_S \frac{\partial u}{\partial n}\, H(Q,P)\, dS - C \iint_S u(Q) dS \tag{64}$$

where H(Q,P) is the Green's function for Neumann problem such that

$$\frac{\partial H}{\partial n} = C \qquad (Q \;\text{ on }\; S) \; . \tag{65}$$

The constant C is determined by following condition that

$$H = \frac{1}{4\pi r} - \Gamma(Q,P)$$

and

$$\iint_S \frac{\partial \Gamma}{\partial n}\, dS = \frac{1}{4\pi} \iint_S \left(\frac{\partial \frac{1}{r}}{\partial n} - 4\pi C\right) dS = 0 \tag{66}$$

It should be noted that

$$-\left[\frac{\partial}{\partial n}\left(\frac{1}{r}\right) dS\right]$$

is the solid angle subtended by dS at P.

<u>EXERCISES</u>

(1) Derive Poisson's integral for a harmonic function in a sphere with the radius R:

$$u(r,\theta,\varphi) = (4\pi R)^{-1} \iint_S f(\theta',\varphi') \frac{R^2 - r^2}{[R^2+r^2-2R\,r\cos\gamma]^{3/2}}\, dS \tag{67}$$

Where S is the surface of the sphere and

$$\cos\gamma = \cos(\theta'-\theta)\sin\varphi\sin\varphi' + \cos\varphi\cos\varphi'$$

(2) Let

$$f(\varphi',\theta') = \frac{\partial u(Q)}{\partial n} \quad \text{for } Q \text{ on } S$$

(where all the notations are the same as in Ex. 1). Show that the harmonic function, with the proper limitation on the derivatives on the boundary, has the representation in the sphere of radius R:

$$u(\rho,\theta,\varphi) = \frac{1}{4\pi} \int_S f(\theta',\varphi') \left[\frac{2}{r} + \frac{1}{R}\log\frac{2R}{R-\rho\cos\gamma + r}\right] dS \tag{68}$$

(3) Show, also, that the integral (68) solves the Neumann problem in a sphere for an arbitrary prescribed continuous function $f(\theta',\varphi')$.

(4) Liouville's theorem

Show, by (67), that if $u \geqq 0$, then

$$u(0)\frac{R(R-r)}{(R+r)^2} \leqq u(P) \leqq \frac{R(R+r)}{R(R-r)^2}\, u(0) . \tag{69}$$

consequently, if u is harmonic in the whole space and is bounded from one-side (say $u \leqq M$), then u is a constant.

HINT: Without the loss of generality, one can assume that $u \geqq 0$. Otherwise consider $v = M-u$ instead of u.

(5) Harnack's First Convergence Theorem of Harmonic Functions

Let $\bar{D}$ be any bounded connected closed domain (regular or not). Show that if the sequence of harmonic functions $u_1, \ldots, u_n$, ... converges uniformly on the boundary S of D, it converges uniformly in $\bar{D}$. The limit $u = \lim_{n\to\infty} u_n$ is harmonic in D. The sequence derived by the partial differentiation converges uniformly to the partial derivatives of in any closed domain in the interior of D.

HINT: i) Apply the fact that the extremum values of a harmonic function lie on the boundary in order to prove the uniform convergence. Consider $(u_{n+p}-u_n)$.

ii) Because of the uniform convergence, one can use Poisson's integral for differentiation.

(6) Harnack's Second Convergence Theorem of Harmonic Functions

If a series of non-negative harmonic functions in a bounded, connected, closed domain $\bar{D}$ converges at one interior point, shown by Ex. 4, that the sequence converges uniformly in any closed domain in the interior of D.

HINT: Consider $u_n+u_{n+1} + \dots + u_{n+p}$, where $u_i \geq 0$.

5.3 Properties of the Green's Function

Let $G(P,Q) = G_d(P,Q)$ be the Green's function for the Dirichlet problem of the domain D with respect to the point P.

Theorem 1.

$$G \geqq 0 .$$

Theorem 2.

$$G(P,Q) = G(Q,P)$$

Theorem 3. If $D' \subset D$, then

$$G_{D'}(P,Q) \leqq G_D(P,Q) \quad \text{for any} \quad P,Q \subset D'$$

The theorem 1 is evident from the properties of the extremum values of a harmonic function. For the proof of the Theorem 2, consider two interior points P_1 and P_2 in D.

Let

$$u = G(P_1,Q) \ , \quad v = G(P_2,Q) \quad .$$

Then the Green second formula gives immediately

$$\iint_{S_1} \left(u \frac{\partial v}{\partial n} - v \frac{\partial u}{\partial n}\right) dS + \iint_{S_2} \left(u \frac{\partial v}{\partial n} - v \frac{\partial u}{\partial n}\right) dS = 0 \qquad (70)$$

where S_1 and S_2 are respectively small spheres with the centers at P_1 and P_2. As S_1 and S_2 shrink to the P_1 and P_2 respectively, the first integral and the second integral in (70) tend respectively to $-G(P_1,P_2)$ and $G(P_2,P_1)$. Thus the symmetrical property of the Green's function with respect to the two points are established.

The theorem 3 follows immediately from the extremum properties of the difference of the Green's functions by means of the theorem 1.

EXERCISES

Show that if $G(a,b;\ x,y)$ is the Green's function of a simply-connected domain D bounded by a Jordan curve P, the function

$$W = e^{g(z)} \tag{71}$$

where

$$Re[g(z)] = 2\pi G(z,b;\ x,b) = \log r - g(a,b;\ x,y)$$

$$z = x + yi\ ,\ \alpha = a + bi\ ,\quad r = |z - \alpha|$$

maps the domain D onto the unit circle. No corresponding theorem exists for the three dimensional case.

6. APPLICATIONS OF THE METHOD OF SOURCES TO THE OTHER TYPES OF PARTIAL DIFFERENTIAL EQUATIONS

Example: Oscillations of an Airfoil in a Supersonic Compressible Fluid.

Consider an infinite cylinder (with the axis along z-axis) moving with a constant speed U in the direction of the negative x-axis. The problem of small transverse oscillations is to be studied. Assume that the cross section is so slender that it is almost a segment of a straight line and coincides with the x-axis in the non-oscillating position. As the case of the vibrating string, the procedure of "linearization" is applied to the general system of hydrodynamical equations (Ex. 1, 2, p. 27, Ch. I). For Eulerian coordinates, let the reference be fixed on the moving airfoil. By the assumption of irrotationality, one may set

$$u = U + \frac{\partial\varphi}{\partial x} \tag{72}$$

$$v = \frac{\partial\varphi}{\partial y} \tag{73}$$

where $\varphi(x,y,t)$ and its derivatives are considered as "small terms." Now φ, defined up to an arbitrary function of t, is called <u>the induced velocity potential</u>. Neglecting the higher order terms in the system of hydrodynamical equations, one obtains

$$(1-M^2)\ \varphi_{xx} + \varphi_{yy} - \frac{1}{a^2}\,(2U\varphi_{xt}+\varphi_{tt}) = 0 \tag{74}$$

where M = the free stream Mach number = $\frac{U}{a}$ and a = the velocity of sound in the still fluid.

The flow is the supersonic if $M > 1$. One needs only to consider the problem for $y \geqq 0$ by symmetry.

(i) The disturbances created at the leading edge propagate along the wedge (called the Mach wedge):

$$y - \frac{x}{(M^2-1)^{\frac{1}{2}}} = 0\ , \qquad (y \geqq 0)\ .$$

This can be explained by considering a moving sound source with uniform velocity in the still fluid as illustrated in the following figure.

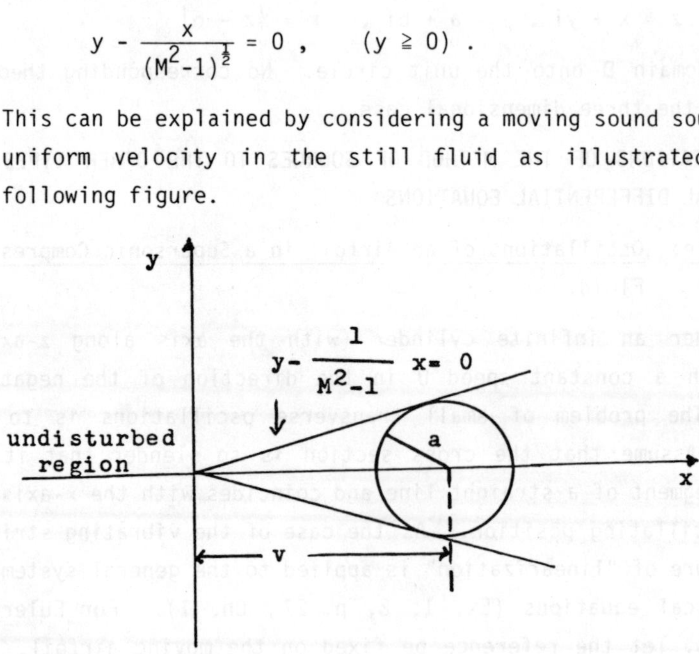

Figure 21

In the front of the Mach wedge, the medium is undisturbed and the flow is therefore uniform relative to the airfoil.

Let the position of the airfoil with respect to the system of moving coordinates be prescribed as:

$$y = G(x,t) \qquad 0 \leqq x \leqq 1$$

The normal component of the velocity of the airfoil is

$$\frac{\partial G}{\partial t} \cos \alpha \ , \quad \text{where} \quad \tan \alpha = \frac{\partial G}{\partial x} \ ,$$

This must be equal to the normal component of the flow $-u \sin\alpha + v \cos \alpha$. Thus the boundary conditions for the present problem are:

$$\text{(i)} \quad \varphi = 0 \ , \quad \text{on} \quad \frac{x}{M^2-1} - y = 0 \tag{75}$$

$$\text{(ii)} \quad \frac{\partial \varphi}{\partial y} = U \frac{\partial G}{\partial x} + \frac{\partial G}{\partial t} + u'(x,y,t) \tan \alpha \tag{76}$$

on the airfoil.

where

$$u = U + u' \ .$$

Within the scope of the linearized theory, the third term in (76) can be neglected. The problem is further simplified by taking this condition on the neutral position (i.e. $y = 0$) instead of the condition on the airfoil. Hence

$$\left. \frac{\partial \varphi}{\partial y} \right|_{y=0} = U \frac{\partial G}{\partial x} + \frac{\partial G}{\partial t}$$

The problem is now to find the solution of the equation (74) with the boundary conditions (75) and (77).

6.1 Possio's Solution

Various methods have been employed by many authors to solve the preceding problem. It was first solved by C, Possio by employing the method of sources.

The Equation (74) can be reduced to the classical two-dimensional wave equation by the following transformation

$$x_1 = x - Ut ,$$

$$y_1 = y , \qquad (78)$$

$$t_1 = t .$$

Thus one elementary solution analogous the source of the Laplace equation is

$$\frac{1}{[a^2t_1^2 - x_1^2 - y_1^2]^{\frac{1}{2}}}$$

As the equation is invariant under a trnaslation of the variables, one has an elementary solution of (74) of the form

$$\varphi_0 = \frac{A(\xi,\eta,\tau)}{[a^2(t-\tau)^2 - \{x-\xi-U(t-\tau)\}^2 - (y-\eta)^2]^{\frac{1}{2}}} \qquad (79)$$

where ξ, η, τ are parameters. This solution may be interpreted physically as the effect at (x,y) due to a sound source of the strength A originating at the point (ξ,η) and at earlier instant τ. As the disturbance is continuously generated at the point (ξ,η), the effect at (x,y) and at t can be tracked back to the disturbances which are generated at (ξ,η) in the time interval (τ_1,τ_2). Because the medium flows with a supersonic speed, the disturbances generating at much earlier instants are carried away to the downstream. The values τ_1 and τ_2 are the roots of the following equation:

$$a^2\tau_i^2 - \{(x-\xi) - U\tau_i\}^2 - (y-\eta)^2 = 0 \qquad (i = 1,2) \qquad (80)$$

Thus the total effect at (x,y) due the disturbances at $(\xi,0)$ is

$$\int_{\tau_1}^{\tau_2} \frac{A(\xi,0,t-\tau)\, d\tau}{[a^2\tau^2 - (x-\xi-U\tau)^2 - y^2]^{\frac{1}{2}}} \qquad (81)$$

Since any disturbance behind the Mach line has no effect in the front, (cf. Fig. 9), one has the solution, from the superposition principle,

$$\varphi(x,y,t) = \frac{1}{a(M^2-1)^{\frac{1}{2}}} \int_0^{\xi_1} d\xi \int_{\tau_1}^{\tau_2} \frac{A(\xi,0,t-\tau)d\tau}{[(\tau-\tau_1)(\tau_2-\tau)]^{\frac{1}{2}}} \tag{82}$$

where

$$\xi_1 = x - (M^2-1)^{\frac{1}{2}} y \tag{83}$$

This form of the solution certainly satisfies the boundary condition (75). That it is a solution of (74) can be rigorously verified. Now it suffices to determine the source intensity A by the condition (77). Set

$$2\tau = (\tau_2-\tau_1) \cos \vartheta + \tau_2 + \tau_1 .$$

One finds readily

$$\varphi = \frac{1}{a(M^2-1)^{\frac{1}{2}}} \int_0^{\xi_1} d\xi \int_0^{\pi} A(\xi,0,t - \frac{\tau_2+\tau_1}{2} - \frac{\tau_2-\tau_1}{2} \cos\theta)\, d\theta \tag{84}$$

Hence,

$$\frac{\partial\varphi}{\partial y} = \frac{1}{a(M^2-1)^{\frac{1}{2}}} \frac{\partial\xi_1}{\partial y} \pi A\, (\xi_1,0,t - \frac{My}{a(M^2-1)^{\frac{1}{2}}})$$

$$+ \frac{-y}{a^2(M^2-1)^{\frac{1}{2}}} \int_0^{\xi_1} d\xi \int_0^{\pi} \frac{\partial A}{\partial t} \cos\theta \frac{1}{[(x-\xi)^2-(M^2-1)y^2]^{\frac{1}{2}}} d\theta \tag{85}$$

As $y \to 0$, by (77), one has

$$-\frac{\pi}{a} A(x,0,t) = \frac{\partial G}{\partial t} + U \frac{\partial G}{\partial x} \tag{86}$$

where G is prescribed. The solution of the problem is then given by (82).

7. POISSON'S EQUATION

Consider the partial differential equation

$$\Delta u = f(x,y,z) \tag{87}$$

where $f(x,y,z)$ is continuously differentiable in the domain $\bar{D}$. Let the boundary condition be such that

$$u = 0 \quad \text{on the boundary} \quad B \quad \text{of} \quad D \tag{88}$$

One may reduce this problem to the Dirichlet problem by finding a particular solution of (87). Under some particular situation, Heavisides operational method may be employed. (cf. Ex. 1).

The solution of the general problem can be expressed as

$$u(P) = \iiint_D G(P,Q)\, f(Q)\, dV_Q \tag{89}$$

where

$$G = \frac{1}{r_{PQ}} - g(P,Q)$$

is the Green's function. Assume that the surface is so smooth that $g(P,Q)$ is twice continuously differentiable in D.

The derivation of this formula is simply an application of the Green's second formula (55) by means of the technique analogous to that in deriving (56). This is left to the reader as an exercise.

The question now arises that whether (89) actually represents the solution of (87) with condition (88). That u, given by (89), satisfies (88) is evident. In order to prove that it satisfies the equation (87), it suffices to establish the following lemma:

Lemma. If $f(\xi,\eta,\zeta)$ is continuous in D then, $u(P)$, given by

$$\iiint_D \frac{1}{r_{PQ}}\, f(Q)\, dV_Q$$

and its first partial derivatives are bounded and continuous in the entire xyz-space. If, in addition,

$f(\xi,\eta,\zeta)$

is continuously differentiable* the second partial

---

*It is seen from the proof that a weaker condition by assuming f to be piecewise continuously differentiable is sufficient.

derivatives u_{xx}, u_{yy}, and u_{zz} exist and are continuous in D and

$$\Delta u = -A\pi f(x,y,z) \quad \text{in D}$$

Proof. Let

$$F(P) = \begin{cases} f(P), & (P \text{ inside } \bar{D}) \\ 0, & (P \text{ outisde } \bar{D}). \end{cases}$$

Let $\delta(P)$ be the sphere of radius ρ with the center at P. Denote the entire space by R and let $R'(P) = R-\delta$. Then

$$u(P) = \iiint_R \frac{F(Q)}{r_{PQ}} dV_Q$$

$$= \iiint_{\delta(P)} \frac{F(Q)}{r_{PQ}} dV_Q + \iiint_{R'(P)} \frac{F(Q)}{r_{PQ}} dV_Q$$

$$= u_1(P) + u_2(P)$$

(i) It is evident that u(P) is continuous in P.

(ii) Denote by P_1 the point $(x+h,y,z)$. Let $r_{PQ} = r$, and

$$r_{P_1Q} = r_1 \quad .$$

Then,

$$|r - r_1| \leqq |h|$$

and

$$|\frac{1}{h}| \cdot |\frac{1}{r} - \frac{1}{r_1}| = |\frac{r_1 - r}{h}| \quad \frac{1}{rr_1} \leqq \frac{1}{rr_1} < (\frac{1}{r^2} + \frac{1}{r_1^2}) \frac{1}{2}$$

Hence

$$|\frac{u_1(P) - u_1(P_1)}{h}| \leqq \frac{1}{2} \iiint_{\delta(P)} |f(Q)| \; (\frac{1}{r^2} + \frac{1}{r_1^2}) \, dV_Q$$

Thus

$$|\frac{u_1(P) - u_1(P_1)}{h}|$$

can be made as small as one pleases provided ρ is small enough. On the other hand,

$$\lim_{h\to 0} \frac{u_1(P_1)-u_2(P)}{h} = -\iiint_{R'} \frac{f(Q)}{r^3}(x-\xi)\, dV_Q$$

Now

$$\left|\frac{x-\xi}{r}\right| \leq 1 .$$

$$\left|\iiint_{\delta(P)} \frac{F(Q)}{r^3}(\xi-x)\, dV_Q\right| \leq 4\pi\rho M .$$

Hence taking h small enough, the expression

$$\left|\frac{u(P_1)-u(P)}{h} + \iiint_D \frac{f(\xi,\eta,\zeta)(x-\xi)}{r^3}\, d\xi\, d\eta\, d\zeta\right|$$

can be made as small as one pleases.

The boundedness and continuity of the partial derivatives can be established exactly by the same reasoning.

(iii) Let P be an interior point. Now, f(Q) is continuously differentiable. Hence

$$u_x(P) = \iiint_D f(Q)\frac{\partial}{\partial x}\left(\frac{1}{r_{PQ}}\right) dV_Q$$

$$= -\iiint_D f(Q)\frac{\partial}{\partial \xi}\left(\frac{1}{r_{PQ}}\right) dV_Q$$

$$= \iiint_D \frac{\partial f(Q)}{\partial \xi}\frac{1}{r_{PQ}}\, dV_Q + \iint_S f(Q)\frac{ds_Q}{r_{PQ}} \tag{90}$$

Where S is the boundary of D. The last integral in (90) is a harmonic function of P. The first integral is continuously differentiable by the same reasoning as in (ii). Consider

$$\Delta u = -4\pi\varphi(x,y,z) .$$

Suppose that

$$f(x_0,y_0,z_0) \neq \varphi(x_0,y_0,z_0)$$

at a point

$$P_0 = (x_0,y_0,z_0)$$

in the interior of D. Assume, without the loss of generality, that $f(P) > \varphi(P)$ for all points in the sphere $\delta(P)$ with P_0 as its center. By the third Green's formula,

$$4\pi \iiint_\delta \frac{\varphi(Q)}{r_{PQ}} dV_Q = 4\pi\, u_1(P) + \iint_S u(Q) \frac{\partial}{\partial n}\left(\frac{1}{r_{PQ}}\right) dS_Q - \iint_S \frac{1}{r_{PQ}} \frac{\partial u(Q)}{\partial n} dS_Q$$

$$= u(P) + \text{q harmonic function in } P.$$

Hence

$$V(P) = \iiint_\delta \frac{f(Q)-\varphi(Q)}{r_{PQ}} dV_Q$$

is a harmonic function inside and outside D and has continuous partial derivatives everywhere. Now

$$\lim_{P\to\infty} V = 0 .$$

Then by the lemma in the following section $V = 0$ which contradicts the fact that $V(P) > 0$.

7.1 Lemma

If V is harmonic and twice continuously differentiable inside and out a domain D and its first partial derivatives are continuous on the boundary S of D, then V is harmonic on S. Consequently*, if

---

*Because a bounded function which is harmonic in the entire space is a constant. (cf. Ex. 4, p. 195).

$$\lim_{P\to\infty} V(P) = 0 ,$$

then

$$V = 0 .$$

Proof. Let P be a point inside the domain D. By the third Green's formula (56)

$$-4\pi V(P) = \iint_{S+S_1} [u(Q) \frac{\partial}{\partial n} (\frac{1}{r_{PQ}}) - \frac{1}{r_{PQ}} \frac{\partial u(Q)}{\partial n}] \, dS_Q \qquad (92)$$

where S_1 is the surface of a sphere inside D (with the point P outside S'). On the other hand, let S_2 be the surface of

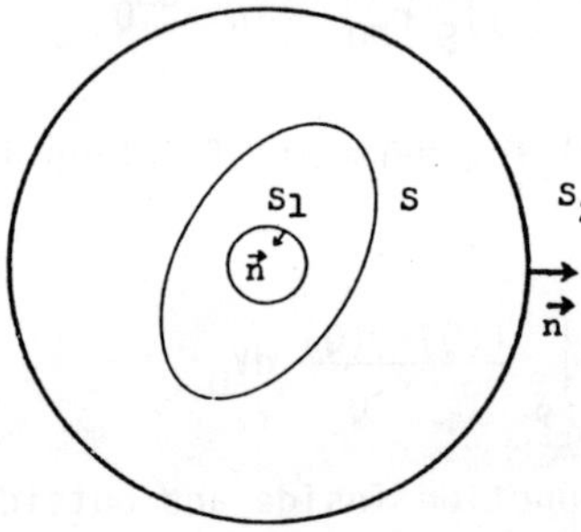

Figure 22

Then

$$0 = \iint_{-S+S_2} [u(Q) \frac{\partial}{\partial n} (\frac{1}{r_{PQ}}) - \frac{1}{r_{PQ}} \frac{\partial}{\partial n} u(Q)] \, dS_Q \qquad (93)$$

The difference (92) and (93) gives the representation

$$-4\pi u(P) = \iint_{S_1+S_2} [u(Q) \frac{\partial}{\partial n} (\frac{1}{r_{PQ}}) - \frac{1}{r_{PQ}} \frac{\partial}{\partial n} u(Q)] \, dS \qquad (94)$$

By the same reasoning, it is true that the representation (94) is valid for the point P in the region bounded by S and S_2. Now (94) represents a harmonic function in domain between S_1 and S_2 on the one hand, it represents u(P) for P on S by continuity on the other hand. Therefore u(P) is harmonic on S. This completes the proof of the last section.

EXERCISES

(1) Solve the following problem

$$\Delta u = e^x (x+y)$$

$$u = 0 \text{ on the unit circle}$$

by

(a) finding a particular solution using the operational method;

(b) reducing it to the Dirichlet problem.

(2) Show that if

(a) $u = -4\pi f(x,y,x)$ in D

(b) $u = 0$, outside D

(c) u and its first partial derivatives on continuous,

and

(d) $\lim_{P\to\infty} u(P) = 0$,

then

$$u = \iiint_D \frac{f(\xi,\eta,\zeta)}{r} \, d\xi \, d\eta \, d\zeta$$

where

$$r = [(x-\xi)^2 + (y-\eta)^2 + (z-\zeta)^2]^{\frac{1}{2}}$$

and f(x,y,z) is continuously differentiable in D.

(3) Simple Layer

Show that if f(Q) is continuous on S, the function defined by

$$u(P) = \iint_S \frac{f(Q)}{r_{PQ}} \, dS_Q$$

possesses the following properties:

(i) Let n_P be the inward normal at a point P_0 on S. $\frac{\partial u}{\partial n_{P_0}}$ exists. Denote the value by $\frac{\partial u(P_0)}{\partial n}$ $\frac{\partial u(P_0)}{\partial n}$

Both

$$\lim_{\substack{P \to P_0 \\ (P \text{ inside } D)}} \frac{\partial u}{\partial n} \quad \text{and} \quad \lim_{\substack{P \to P_0 \\ (P \text{ outside } D)}} \frac{\partial u}{\partial n} \quad \text{exists.}$$

Denote them respectively by

$$\frac{\partial u+(P_0)}{\partial n} \quad \text{and} \quad \frac{\partial u-(P_0)}{\partial n}$$

$$\frac{\partial u+(P_0)}{\partial n} - \frac{\partial u(P_0)}{\partial n} = -2\pi f(P_0)$$

$$\frac{\partial u(P_0)}{\partial n} - \frac{\partial u-(P_0)}{\partial n} = -2\pi f(P_0)$$

(4) <u>Double Layer</u>

Show that if f(Q) is continuous on S, then the function defined by

$$u(P) = \iint_S f(Q) \frac{\partial}{\partial n_Q} \left(\frac{1}{r_{PQ}}\right) dS_Q$$

has a discontinuity at a point P_0 on S; namely

$$u^+(P_0) - u(P_0) = 2\pi f(P_0)$$

$$u^-(P_0) - u(P_0) = 2\pi f(P_0)$$

where

$$u^+(P_0) = \lim_{\substack{P \to P_0 \\ (P \text{ inside } D)}} u(P_0)$$

and

$$u^-(P_0) = \lim_{\substack{P \to P_0 \\ (P \text{ outside } D)}} u(P_0)$$

(5) Establish the theorems corresponding to Ex. 3 and Ex. 4 for the two-dimensional case.

(6) Reduction of the Dirichlet Problem to an Integral Equation

Let Γ be a regular closed (Jordan) curve with the continuous tangent. Let D be the domain bounded by Γ. Analytically, Γ can be represented by

$$\begin{aligned} x &= \xi(s) \\ y &= \eta(s) \end{aligned} \tag{95}$$

$(0 \leqq s \leqq a)$ such that $\xi(0) = \xi(a)$ and $\eta(0) = \eta(a)$. Show that the Dirichlet problem in D has a solution $u(x,y)$ if there exists $\mu(s)$ such that

$$u(x,y) = \int_0^a \frac{1}{2}\,\mu(s)\,\frac{\partial}{\partial n}\log\frac{1}{\{x-\xi(s)\}^2 + \{y-\eta(s)\}^2}\,dP$$

where $\mu(s)$ is continuous on Γ such that

$$u^-(\xi(s)\,,\ \eta(s)) = f(s)$$

and $f(s)$ is prescribed value of u on Γ. In other words, $\mu(s)$ satisfies the integral equation

$$\mu(\sigma) = f(\sigma) - \int_0^a K(\sigma,s)\,\mu(s)\,ds \tag{96}$$

where

$$K(\sigma,s) = \frac{1}{\pi}\,\frac{\{\eta(\sigma)-\eta(s)\}\xi'(s) - \{\xi(\psi)-\xi(s)\}\,\eta'(s)}{\{\xi(\sigma)-\xi(s)\}^2 + \{\eta(\sigma)-\eta(s)\}^2} \tag{97}$$

The existence of the solution (96) is established in the theory of integral equations. It is remarked that the existence of the solution of the Dirichlet problem for the more general domain has been a subject of the very extensive study.

7.2 Steady Compressible Flows around an Airfoil

Let $\varphi^*(x,y)$ be the velocity potential; namely,

$$u = \varphi^*_x$$
$$v = \varphi^*_y \tag{98}$$

where u and v are respectively the x-component and the y-component of the flow velocity. The fundamental hydrodynamical equations (Ex. 1 and Ex. 2 on p. 50) give the following equation

$$(a^2-u^2)\,\varphi^*_{xx} - 2uv\,\varphi^*_{xy} + (a^2-y^2)\,\varphi^*_{yy} = 0 \tag{99}$$

where

$$a^2 = \gamma\,\frac{p}{\rho} = a_0^2 - \frac{\gamma-1}{2}\,(U^2-u^2-v^2) \tag{100}$$

a_0 = the velocity of sound of the free stream

U = the speed of the free stream

Put $\varphi^* = U$. The equation (99) can be rewritten as

$$\varphi_{xx} + \varphi_{yy} = \frac{M_0^2}{1-\frac{\gamma-1}{2}M_0^2}\left[\left(\frac{\gamma+1}{2}\,\varphi_x^2 + \frac{\gamma-1}{2}\,\varphi_y^2\right)\varphi_{xx} + 2\varphi_x\varphi_y\varphi_{xy}\right.$$
$$\left. + \left(\frac{\gamma-1}{2}\,\varphi_x^2 + \frac{\gamma+1}{2}\,\varphi_y^2\right)\varphi_{yy}\right]$$
$$= k\,L(\varphi) \tag{101}$$

where

$$k = \frac{M_0^2}{1-\frac{\gamma-1}{2}M_0^2}$$

$M_0 = \frac{U}{a_0}$ = the free stream Mach number

The boundary conditions are

(i) $\frac{\partial\varphi}{\partial n} = 0$ on the airfoil (102a)

(ii) $\varphi_x \to 1$ and $\varphi_y \to 0$ as $(x^2+y^2) \to \infty$; (102b)

and

(iii) The Kutta-Joukowsky condition - the flow velocity is finite everywhere in the fluid and on the airfoil Γ. (102c)

Besides, one assumes that the derivatives of φ are single-valued.

7.2.1 The Method of Perturbation Based on a Small Parameter. The Rayleigh-Janzen's Solution. For the boundary-value problem of a non-linear differential equation, there is so far no general method available in obtaining the exact solution of the problem. In the case of a small parameter, the method of perturbation is very often proved to be effective. The following method of solving the preceding problem is due to Rayleigh and Janzen*. Let

$$\varphi(x,y) = \varphi_0(x,y) + k\,\varphi_1(x,y) + k^2\,\varphi_2(x,y) + \dots \tag{103}$$

Substituting in (101) and equating the coefficients of the powers of k, one has

$$\Delta\varphi_0 = 0$$

Here φ_0 satisfies all the boundary conditions (102). Thus φ_0 is the solution for incompressible flow. Next,

$$\Delta\phi_1 = F_1(\varphi_0) \tag{104}$$

and in general

$$\phi_n = F_n(\varphi_0, \varphi_1, \dots \varphi_{n-1}) \tag{105}$$

The boundary conditions for φ_1, are (102a), (102c) and the first partial derivatives tend to zero as $x^2+y^2 \to \infty$. The same boundary

---

*Rayleigh and Janzen gave independently the form of the solution without a rigorous justification. The mathematical justification is due of F. Frankl and M. Keldysch, Die Aussere Neumann'sche Aufgabe fur nicht-lineare ellisptische Differentialglechungen mit auf die theorie des Flugels im kompressiblen Gas, Bull. Acad. Sc., U.S.S.R., pp. 561-601 (1934).

conditions are set up for φ_n ($n \geqq 1$). Besides, the derivatives of φ_n are assumed to be single-valued. Thus, one can determine the functions φ_n successively by solving the Poisson's equation with homogeneous boundary conditions.

That, for small values of M_0, the series (103) actually converges and represents the solution of the problem is due Frankl and Keldvsch.

EXERCISE

Solve the problem of compressible flow around an infinite circular cylinder by above method. Assume that the circulation

$$\oint_{\Gamma^*} d\varphi = 0$$

where Γ^* is any simple closed curve containing Γ in its interior.

8. THE GREEN'S FUNCTION OF THE BIHARMONIC EQUATION

The Clamped Plate with a Concentrated Load.

Consider the partial differential equation

$$\Delta\Delta w = \frac{q(x,y)}{k} \tag{105}$$

with the boundary conditions:

$$w \equiv 0 \quad \text{and} \quad \frac{\partial w}{\partial n} = 0 \tag{106}$$

on the boundary Γ of the domain D. One may interpret w as the transverse displacement of an isotropic thin plate. $q(x,y)$ is the load per unit area and the constant k is the flexural rigidity of the plate. In the case of a concentrated load; namely Γ is a δ-function at a point $Q(\xi,\eta)$, the solution is called the Green's function. The precise meaning is given as follows:

Definition. If for a fixed point Q in D, a function G(P,Q) exists such that

(i) $\Delta\Delta_P G = 0$ in D except at $P = Q$,

(ii) $G = 0$ and $\dfrac{\partial G}{\partial n_P} = 0$ on Γ,

and

(iii) $G = \frac{1}{8\pi} r_{PQ}^2 \log r_{PQ} - \gamma(P,Q)$

where r_{PQ} = the distance between P and Q and $\gamma(P,Q)$ is a biharmonic function in D, then G(P,Q) is defined as the Green's function of the domain D at the point Q. Once the Green's function exists, the solution of (105) with the boundary conditions (106) is given by the following formula:

$$u(Q) = \iint_D G(P,Q) \frac{q(P)}{k} \, dA_P \tag{107}$$

The derivation of (107) is based on the Rayleigh-Green's formula:

$$\iint_D (u \cdot \Delta\Delta v - v \cdot \Delta\Delta u) \, dA$$

$$= \int_\Gamma \left(u \frac{\partial \Delta v}{\partial n} - \Delta v \frac{\partial u}{\partial n} - v \frac{\partial \Delta u}{\partial n} + \Delta u \frac{\partial v}{\partial n}\right) dS \tag{108}$$

where n denotes the outward normal. The general form of Rayleigh-Green's formula is given in Ex. 1.
The boundary value problem

$$\Delta\Delta u = 0$$

Such that u and $\frac{\partial u}{\partial n}$ are prescribed, say, a and b on the boundary Γ of D can also be solved by means of the Green's function. The method is an analogy to the first boundary problem of the harmonic function. One has

$$u(Q) = \int_\Gamma \left(a \frac{\partial \Delta G}{\partial n} - b \Delta G\right) ds_P \tag{109}$$

<u>EXERCISES</u>

(1) Establish the Rayleigh-Green's formula

$$\iint_D (u\Delta\Delta v - v\cdot\Delta\Delta u)\, dA$$

$$= \int_\Gamma \left(uI_3(v) - \frac{\partial u}{\partial n} I_2(v) + \frac{\partial v}{\partial n} I_2(u) - vI_3(u)\right) ds \quad (110)$$

where

$\vec{n} = (n_1, n_2)$ is a unit vector along the inward normal

$$I_2(u) = \mu\Delta u + (1-\mu)\,(n_1^2 u_{xx} + 2n_1 n_2 u_{xy} + n_2^2 u_{yy})$$

$$I_3(u) = \frac{\partial \Delta u}{\partial n} - (1-\mu)\frac{\partial}{\partial s}\,[n_1 n_2\,(u_{yy}-u_{xx}) + (n_1^2 - n_2^2)\, u_{xy}]$$

where $\frac{\partial u}{\partial s}$ is the tangential derivative along the boundary (in the positive sense) and μ is a constant.

<u>HINT</u>:

(i) Establish (108) by the second Green's formula.

(ii) The formula (110) is established if one proves that

$$\int_\Gamma [u\, I_3'(v) - vI_3'(u) - \frac{\partial u}{\partial n} I_2'(v) + \frac{\partial v}{\partial n} I_2'(u)]\, ds = 0$$

where

$$I_2'(u) = -\Delta u + n_1^2\, u_{xx} + 2n_1 n_2 u_{xy} + n_2^2\, u_{yy}$$

$$I_3'(u) = -\frac{\partial}{\partial s}\,[n_1 n_2\,(u_{yy} - u_{xx}) + (n_1^2 - n_2^2)\, u_{xy}]$$

(iii) Integrate by parts of the terms involving I'. Show that the integrand is identically zero. Use the fact that $\vec{n} = (n_1, n_2)$, $\vec{s} = (n_2, -n_1)$.

(2) Derive the formula corresponding to (109) for the following problem:

$$\Delta\Delta u = 0$$

with the boundary conditions on Γ:

$$u = a$$

$$I_2(u) = b$$

8.1 Application of Residue Theory to Boundary Value Problem of Biharmonic Equation in a Circle

The residue theory is especially useful in attacking the circle problem. Consider, for instance, the biharmonic function on u such that

$$\Delta\Delta u = 0 \quad \text{in} \quad D: \; r^2 = x^2 + y^2 < 1 \, . \tag{111}$$

Let the boundary values be

$$u(1,\theta) = a(\theta) \, , \tag{112}$$

$$u_r(1,\theta) = b(\theta) \, . \tag{113}$$

First of all, it should be remarked that a single-valued biharmonic function in any simply-connected domain of existence, has the representation

$$u = \mathrm{Re}\,[\bar{z}\, f(z) + g(z)] \, . \tag{114}$$

where $f(z)$ and $g(z)$ are two analytic functions in the domain and $z = x+yi$. Let

$$\Delta u = p(x,y)$$

and

$$4f'(z) = p(x,y) + iq(x,y)$$

where $q(x,y)$ is defined up to a real constant. Let

$$f(z) = P(x,y) + iQ(x,y)$$

then

$$\Delta(u-xP-yQ) = p - 2\frac{\partial P}{\partial x} - 2\frac{\partial Q}{\partial y}$$

$$= p - 4\frac{\partial P}{\partial x} = 0 \, .$$

Hence u - xP - yQ is a single-valued harmonic function and thus the representation is established.

Theorem. The solution of the equation (111) with the boundary values (112) and (113) is

$$u(r,\theta) = \mathrm{Re}\,\Big[\frac{1}{2\pi i}\oint_{\Gamma} a(\varphi)\,\frac{z+\alpha}{z(z-\alpha)}\,dz + \frac{\alpha\bar{\alpha}-1}{4\pi i}\oint_{\Gamma}\Big(b(\varphi)\,\frac{z+\alpha}{z(z-\alpha)}-2a(\varphi)\,\frac{\alpha}{(z-\alpha)^2}\Big)\,dz\Big]$$

$$= \frac{1-r^2}{2\pi}\int_0^{2\pi}\frac{a(\varphi)+\frac{1}{2}(r^2-1)b(\varphi)}{1+r^2-2r\cos(\varphi-\theta)}\,d\varphi - \frac{(r^2-1)}{2\pi}\int_0^{2\pi}\frac{a(\varphi)\cos\{(\varphi-\theta)+r^2\cos(\varphi-\theta)-2r\}}{\{1+r^2-2r\cos(\varphi-\theta)\}^2}\,d\varphi \tag{115}$$

($z = e^{\varphi i}$, $\alpha = re^{\theta i}$). Any function which is 4-times continuously differentiable in the circle, and continuously differentiable in the closed circle and which satisfies the biharmonic equation can be represented by the formula (115). Consequently, any two such biharmonic functions must be identically equal if they, together with normal derivatives, are equal on the circle.

The formula can be easily extended to any circle (cf. Ex. 1). The proof given below is based on the additional assumption that the function is biharmonic on the circle. This condition is easily removed by a limiting process; namely, the formula is established for an interior circle with the same center and then let it tend to the circle in consideration.

Proof. By means of (114), the boundary conditions give

$$\frac{f(z)}{z} + z\,\overline{f(z)} + g(z) + \overline{g(z)} = 2a(\varphi) \tag{116}$$

$$\frac{f(z)}{z} + z\,\overline{f(z)} + f'(z) + \overline{f'(z)} + g'(z) + \overline{g'(z)} = 2b(\varphi) \quad (117)$$

where $z = e^{\varphi i}$. Let α be a point inside the unit circle and different from zero. Multiplying (116) by $\frac{1}{z-\alpha}$ and integrating over the unit circle $|z| = 1$, it follows that

$$\frac{f(\alpha)}{\alpha} - \frac{f(0)}{\alpha} + \alpha\overline{f(0)} + \overline{f'(0)} + g(\alpha) + \overline{g(0)} = \frac{1}{\pi i}\oint_\Gamma \frac{a(\varphi)}{z-\alpha}\,dz \quad (118)$$

As $\alpha \to 0$,

$$f'(\alpha) + \overline{f'(0)} + g(0) + \overline{g(0)} = \frac{1}{2\pi i}\oint_\Gamma \frac{a(\varphi)}{z}\,dz \quad (119)$$

From (118) and (119), one has

$$\frac{f(\alpha)}{\alpha} - \frac{f(0)}{\alpha} + \alpha\overline{f(0)} + g(\alpha) = \frac{1}{2\pi i}\oint_\Gamma \frac{a(\varphi)(z+\alpha)}{z(z-\alpha)}\,dz + c_1 i \quad (120)$$

where c_1 is a real constant. Similarly from (117), one obtains

$$\frac{f(\alpha)}{\alpha} - \frac{f(0)}{\alpha} + \alpha\overline{f(0)} + f'(\alpha) + \alpha g'(\alpha) = \frac{1}{2\pi i}\oint_\Gamma \frac{b(\varphi)(z+\alpha)}{z(z-\alpha)}\,dz + c_2 i \quad (121)$$

(c_2 = a real constant).

Differentiating (120) by α and multiplying by α, it gives

$$f'(\alpha) - \frac{f(\alpha)}{\alpha} + \frac{f(0)}{\alpha} + \alpha\overline{f(0)} + \alpha g'(\alpha) = \frac{\alpha}{\pi i}\oint_\Gamma \frac{a(\varphi)}{(z-\alpha)^2}\,dz \quad (122)$$

The equations (121) and (122) give

$$f(\alpha) - f(0) = \frac{\alpha}{4\pi i}\oint_\Gamma \frac{b(\varphi)(z+\alpha)}{z(z-\alpha)} - \frac{2\alpha\, a(\varphi)}{(z-\alpha)^2}\,dz + \frac{\alpha}{2}c_1 i \quad (123)$$

and

$$g(\alpha) = \frac{1}{2\pi i}\oint_\Gamma \frac{a(\varphi)(z+\alpha)}{z(z-\alpha)}\,dz + (c_1 - \tfrac{1}{2}c_2)\,i - \alpha\overline{f(0)}$$

$$- \frac{1}{4\pi i}\oint_\Gamma \left[\frac{b(\varphi)(z+\alpha)}{z(z-\alpha)} - \frac{2\alpha\, a(\varphi)}{(z-\alpha)^2}\right] dz \quad (124)$$

The last two equations give immediately the expression (115). This completes the proof.

EXERCISES

(1) Show that the solution of the equation (111) in a circle of the radius r_0 with the boundary values (112) and (113) is

$$u(r,\theta) = \frac{r_0^2-r^2}{2\pi}\int_0^{2\pi}\frac{a(\varphi)+\frac{1}{2r_0}\,b(\varphi)(r^2-r_0^2)}{r_0^2+r^2-2rr_0\cos(\varphi-\theta)}\,d\varphi$$

$$-\frac{(r^2-r_0^2)}{2\pi r_0}\int_0^{2\pi}\frac{a(\varphi)\{r_0^2\cos(\varphi-\theta)+r^2\cos(\varphi-\theta)-2rr_0\}}{\{r_0^2+r^2-2rr_0\cos(\varphi-\theta)\}^2}\,d\varphi \tag{125}$$

(2) Show that, in (115), if $a(\varphi)$ and $b(\varphi)$ are continuous and of the period 2π, then

$$\lim_{\substack{r\to1\\ \theta\to\theta_0}} u(r,\theta) = a(\theta_0) \tag{126}$$

and

$$\lim_{\substack{r\to1\\ \theta\to\theta_0}} u_r(r,\theta) = b(\theta_0) \tag{127}$$

HINT: Use the technique of Ex. 2 of Section 2.3, p. 167.

Consider the integral of the form

$$\int_0^{2\pi} k(r,\theta,\varphi)\,[c(\varphi)-c(\theta_0)]\,d\varphi$$

where $c(\varphi)$ is either $a(\varphi)$ or $b(\varphi)$. Show that the integral

$$\int_{\theta_0-\delta}^{\theta_0+\delta} k[c(\varphi)-c(\theta_0)]\,d\varphi$$

can be made as small as one pleases for all values of θ in

$(\theta_0 - \frac{\delta}{2}, \ \theta_0 + \frac{\delta}{2})$ and all values of r in consideration provided δ is small enough.

(i) (126) is evident if the terms with the factor (r^2-1) are shown to tend to zero as $r \to 1$ and $\theta \to \theta_0$. Observe that

$$(1+r^2)\cos(\theta-\varphi) - 2r = [1+\cos(\theta-\varphi)](1-r)^2$$

$$- [1+r^2-2r\cos(\theta-\varphi)] \quad .$$

(ii) Show that

$$\frac{(1-r^2)^2}{2\pi}\int_0^{2\pi} \frac{[b(\varphi)-b(\theta)]\,[\cos(\theta-\phi)-r]}{[1+r^2-2r\cos(\theta-\varphi)]^2} d\varphi \to 0$$

as $r \to 1$ and $\theta \to \theta_0$. Observe that

$$\cos(\theta-\varphi)-r = \frac{1}{2r}\,[-\{1+r^2-2r\cos(\varphi-\theta)\} + 1 - r^2]$$

and

$$\frac{1}{2\pi}\int_0^{2\pi} \frac{1}{[1+r^2-2r\cos(\theta-\varphi)]^2}\, d\varphi = \frac{1}{2\pi i}\oint_\Gamma \frac{z\,dz}{(z-\alpha)^2(1-\alpha z)^2}$$

$$= \frac{1+r^2}{(1-r^2)^3}$$

where

$$\alpha = re^{i\theta} \quad , \qquad z = e^{\varphi i} \quad .$$

(iii) Show that when $b(\varphi) = 0$,

$$v = \frac{\partial u}{\partial r} = \mathrm{Re}\Big[\frac{1}{2\pi i}\,\frac{\alpha}{r}\,(1-r^2)\oint_\Gamma \frac{3(z-\alpha)+2\alpha}{(z-\alpha)^3}\,\{a(\varphi)-a(\theta_0)\}\,d\varphi\Big]$$

$\to 0$ as $r \to 1$ and $\theta \to \theta_0$. The first integral tends to zero as shown above. The second integral is

$$\frac{(1-r^2)r}{2}\int_0^{2\pi}\frac{2\{\cos 2(\theta-\varphi)+3r^2-(3r+r^3)\cos(\theta-\varphi)\}}{\{1+r^2-2r\cos(\theta-\varphi)\}^3}\{a(\varphi)-a(\theta_0)\,d\varphi.$$

Observe that

$$2\chi^2-(3r+r^3)\chi+3r^2-1$$

$$=\frac{(1+r)(1-r)^3}{r}(1+\chi)+\frac{1}{r}(1+r^2-2r\chi)(r^2-1+r-\chi)$$

$$=\frac{(1+r)(1-r)^3}{r}(1+\chi)+\frac{1}{r}(1+r^2-2r\chi)^2$$

$$+\frac{1}{r}(1+r^2-2r\chi)[-2+r+(-1+2r)\chi]$$

The last factor

$$-2+r+(-1+2r)\chi \leqq 0 \quad \text{for} \quad |\chi|\leqq 1 \ .$$

(3) Obtain the formula analogous to (115) for the following boundary value problem in the unit circle:

$$\Delta\Delta u = 0$$

$$u(1,\theta) = a(\theta)$$

$$I_2u(1,\theta) = \mu u + (1-\mu)\left.\frac{\partial^2 u}{\partial r^2}\right|_{r=1} = b(\theta)$$

8.2 The Solution of a Concentrated Load of a Circular Clamped Plate

As an application of the preceding discussions, the solution of a concentration load on a circular clamped plate is explicitly given.*

---

*The solution is usually given as an infinite series. (cf., for instance, Timoshenko, theory of Plates and Shells, pp. 266-269). Here the solution is worked out in a closed form. Cf., also M.M. Fridman, The Bending of a Circular Plate by Concentrated Forces, Akad. Nauk. SSSR. Prikl. Mat. Meh. 15, pp. 258-260(1951). The above method is originally due to H. Reissner.

Let β be a fixed interior point in the unit circle on the complex plane. It is to find the biharmonic function $g(z;\beta)$ such that when $z = (re^{\varphi i})$ is on the unit circle (i.e., $r = 1$),

$$g(z,\beta)\Big|_{r=1} = (\beta-z)(\bar{\beta}-\bar{z}) \log (\beta-z)(\bar{\beta}-\bar{z})$$

$$= (\beta\bar{\beta}+1-\bar{\beta}z - \frac{\beta}{z}) \log (1 - \frac{\beta}{z})(1-\bar{\beta}z) = a(\varphi)$$

$$\frac{\partial g(z,\beta)}{\partial r}\Big|_{r=1} = (2-\bar{\beta}z - \frac{\beta}{z}) [\log (1 - \frac{\beta}{z})(1-\bar{\beta}z) + 1] = b(\varphi)$$

One may, now evaluate the integrals in (115) by means of the contour integration. All can be reduced to calculate the integrals of the following types

$$\text{(i)} \quad \frac{1}{2\pi i} \oint_\Gamma \frac{1}{(z-\alpha)^n} \log \frac{(z-\beta)(1-\bar{\beta}z)}{z} dz ,$$

where Γ is the unit circle $|z| = 1$ and $n = 1$ or 2. For the case $\alpha \neq 0$, the contour integration can be evaluated by defining the function

$$\log (1 - \frac{\beta}{z}) (1-\bar{\beta}z)$$

as a single-valued function in the unit circle except on the cut from 0 to β.

$$\text{(ii)} \quad \frac{1}{2\pi i} \oint \frac{1}{z^n} \log (1- \frac{\beta}{z})(1-\bar{\beta}z) dz$$

where $n = 0;$ 1 or 2.

This integral can be evlauated directly by the series expansions:

$$\log (1- \frac{\beta}{z}) = \log (1-\beta e^{-\varphi i}) = -\beta e^{-\varphi i} - \frac{\beta^2 e^{-2\varphi i}}{2} - \ldots$$

$$\log (1-\bar{\beta}z) = -\bar{\beta}\, e^{\varphi i} - \frac{\beta^2 e^{2\varphi i}}{2} - \ldots -$$

It is then a straight forward calculation. The result which can be verified directly is

$$g(\alpha,\beta) = \mathrm{Re}\Big[\alpha\bar{\alpha}\,\Big(1 - \frac{\beta}{\alpha}\Big)\log\,(1-\bar{\beta}\alpha) + (-\alpha\bar{\beta}+\beta\bar{\beta})\,\log\,(1-\bar{\beta}\alpha)$$

$$+ \frac{\alpha\bar{\alpha}-1}{2}\,(-\,\beta\bar{\beta}+\alpha\bar{\beta}+2-2\alpha\bar{\beta})\ .$$

APPENDIX

THE OPERATIONAL METHOD (II)

Cauchy's Residue Theory

The theory of the complex functions is widely used in solving the boundary value problems. It also plays an important role to obtain the inverse Laplace transform from the inversion formula. A brief account is given here in order to facilitate the treatment in the present text. It is intended for those who do not have any previous knowledge of the subject.

1. GEOMETRICAL REPRESENTATION OF A COMPLEX NUMBER

The reader is assumed to be familiar with the notion of the complex numbers. Let $z = x+yi$ where both x and y are real quantities. The pair (x,y) enables one to identify a point on xy-plane with the complex number z. Let (r,θ) be the polar coordinates of the point. The following nomenclature and notations are generally used:

$$r = \text{the modulus or the absolute value of } z = |z| = (x^2+y^2)^{\frac{1}{2}}$$

$$\theta = \text{the argument of } z = \arg(z) .$$

One can readily derive the following rules:

$$|z_1 \cdot z_2| = |z_1| \cdot |z_2| \tag{1}$$

$$\left|\frac{z_1}{z_2}\right| = |z_1| \ / \ |z_2| \tag{2}$$

$$|z_1+z_2| \leqq |z_1| + |z_2| \tag{3}$$

or

$$|z_1-z_2| \geqq \Big||z_1| - |z_2|\Big| \tag{3'}$$

The inequality (3) follows from the fact that a complex number can also be interpreted as a vector from the origin to the point corresponding to it. It is readily verified that the sum of the

components of two vectors is equal to the component of the vector sum. Then (3) is merely a rewording of the triangular inequality. Such geometrical interpretations are of paramount importance. For example, $|z-a| = R$ means the circle with center at a and the radius R. Also, $|z-a| \leqq R$ represents all the points inside the circle.
A generalization of (3) is:

$$|z_1 + z_2 + \ldots + z_n| \leqq |z_1| + \ldots + |z_n| \tag{4}$$

It should be remarked that $\theta = \arg z$ is defined up to an integral multiple of 2π. The following rules are evident.

$$\text{Arg } (z_1 \cdot z_2) = \text{Arg } z_1 + \text{Arg } z_2 \tag{5}$$

$$\text{Arg } (z_1/z_2) = \text{Arg } z_1 - \text{Arg } z_2 \tag{6}$$

The reflection with respect to the x-axis (or the real axis) on the z-plane gives also an important notion. $z = x-yi$ is defined as the complex conjugate of z. It is true that:

$$\overline{z_1 \circ z_2} = \overline{z_1} \circ \overline{z_2} \tag{7}$$

where the operation $\circ$ stands for anyone of the rational operations +, -, x and ÷.

2. FUNCTIONS, ANALYTIC FUNCTIONS

The notions of the function of a complex variable z, the limit and the continuity are defined exactly the same way as in the calculus.

Definition. A function f(z) of the complex variable $z = x+yi$ is said to be analytic at z if the limit of the quotient

$$\frac{f(x+\Delta z) - f(z)}{\Delta z}$$

as $\Delta z = \Delta x+\Delta yi$ tends to 0. The limiting value, if finite, is denoted by $f'(z)$ or $\frac{df(z)}{dz}$.

As the definition is, formally, the same as the derivative of a function of a real variable being defined in the calculus, the following rules are established by the same reasoning:

$$\frac{d}{dz}\{f(z)+g(z)\} = f'(z) \pm g'(z) \tag{8}$$

$$\frac{d}{dz}\{f(z)\cdot g(z)\} = f'(z)\, g(z) + f(z)g'(z) \tag{9}$$

$$\frac{d}{dz}\{\frac{1}{f(z)}\} = -\frac{1}{f^2(z)} f'(z)\,, \quad \text{provided } f(z) \neq 0 \tag{10}$$

If:

$$w = f(u) \quad \text{and} \quad u = g(z)\,,$$

then

$$\frac{dw}{dz} = f'(u)\, g'(z) \tag{11}$$

In all these formulas, the derivatives, wherever they appear, are assumed to exist.

2.1 Cauchy-Riemann Differential Equations

Let:

$$f(z) = P(x,y) + iQ(x,y)$$

where P and Q are both real-valued. Assume that P and Q, together with their first partial derivatives, are continuous at the point z.

Theorem 1. A necessary and sufficient condition that the function is analytic at a point z is that P and Q satisfy the Cauchy-Riemann differential equations:

$$\frac{\partial P}{\partial x} = \frac{\partial Q}{\partial y} \tag{12}$$

$$\frac{\partial P}{\partial y} = -\frac{\partial Q}{\partial x} \tag{13}$$

Remark. These equations can be remembered by the aid of the following rule as indicated in the diagram.

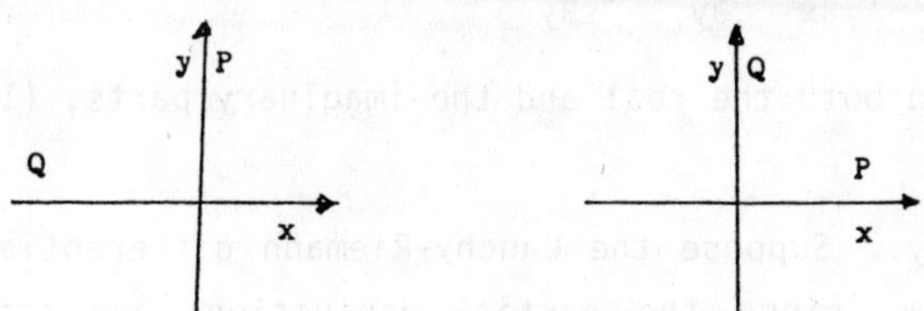

Figure 23

The Cauchy-Riemann differential equations in polar coordinates are:

$$\frac{\partial P}{\partial r} = \frac{1}{r}\frac{\partial Q}{\partial \theta} \tag{14}$$

$$\frac{1}{r}\frac{\partial P}{\partial \theta} = -\frac{\partial Q}{\partial r} \tag{15}$$

which are immediately obtained from (12) and (13) by transformation and which can be remembered in the same way.

<u>Proof of the Theorem</u>

(A) <u>Necessity</u>. Assume the analyticity at z. Then, the limit of quotient:

$$\frac{\Delta f}{\Delta z}$$

exists as $\Delta z \to 0$. One just considers two particular ways for which z tends to zero; namely:

(A_1) $\Delta z = \Delta x \to 0$

(A_2) $\Delta z = i\Delta y \to 0$

by (A_1),

$$\frac{\Delta f}{\Delta z} = \frac{\Delta P}{\Delta x} + i\,\frac{\Delta Q}{\Delta x} \to \frac{\partial P}{\partial x} + i\,\frac{\partial Q}{\partial x}$$

By (A_2),

$$\frac{\Delta f}{\Delta z} = \frac{\Delta P}{i\Delta y} + \frac{\Delta Q}{\Delta y} \to -i\,\frac{\partial P}{\partial y} + \frac{\partial Q}{\partial y}$$

Since the limit must be the same, one concludes that:

$$\frac{\partial P}{\partial x} + i\,\frac{\partial Q}{\partial x} = \frac{\partial Q}{\partial y} - i\,\frac{\partial P}{\partial y}$$

By equating both the real and the imaginary parts, (12) and (13) follow.

(B) <u>Sufficiency</u>. Suppose the Cauchy-Riemann differential equations hold. Now, since the partial derivations are assumed to be continuous at (x,y), one has, by the law of the mean, that:

$$\Delta f(x) = \Delta P + i\Delta Q$$

where

$$\Delta P = P(x+\Delta x,\ y+\Delta y) - P(x,y)$$

$$= P_x(x,y)\Delta x + P_y(x,y)\Delta y + \varepsilon_1\Delta x + \varepsilon_2\Delta y$$

$$\Delta Q = Q_x(x,y)\Delta x + Q_y(x,y)\Delta y + \varepsilon_3\Delta x + \varepsilon_4\Delta y \ .$$

$$= -P_y\Delta x + P_x\Delta y + \varepsilon_3\Delta x + \varepsilon_4\Delta y$$

and

$$\varepsilon_j(j = 1,2,3,4) \to 0 \quad \text{as} \quad \Delta z \to 0 \ .$$

Thus:

$$\frac{\Delta f}{\Delta x} = P_x - iP_y + (\varepsilon_1+i\varepsilon_3)\frac{\Delta x}{\Delta z} + (\varepsilon_2+i\varepsilon_4)\frac{\Delta y}{\Delta z}$$

Since:

$$\left|\frac{\Delta x}{\Delta z}\right| \ , \ \left|\frac{\Delta y}{\Delta z}\right| \leq 1 \ ,$$

it is clear that:

$$\lim_{\Delta z\to 0}\frac{\Delta f}{\Delta z}$$

exists. This completes the proof.

<u>EXERCISES</u>

(1) Let e^z be defined by the following infinite series:

$$1 + z + \frac{z^2}{2!} + \ldots + \frac{z^m}{m!} + \ldots$$

which converges for all complex values of z. Show that:

(a) $e^{yi} = \cos y + i \sin y \qquad (y = \text{real})$

(b) $e^{z_1} \cdot e^{z_2} = e^{z_1+z_2}$

(c) $e^{z+k} = e^z$ for all z if and only if:

$$k = 2n\pi i$$

where n is an integer.

(d) By means of the Cauchy-Riemann differential equations in polar coordinates, e^z is analytic everywhere in the finite plane and:

$$\frac{d}{dz} e^z = e^z \quad ,$$

(2) Let $w = z^n$ (where n is any complex number) be defined as follows:

$$r^n = e^{n \log r}$$

$$w = r^n \, e^{(\theta+2k\pi)\, ni}$$

where k is a fixed integer. The function is, in general, many-valued. But if a cut (say, the positive real axis including the origin)* is excluded from the whole z-plane, show that the function is analytic at every finite value z (not on the cut) and:

$$\frac{d}{dz} z^n = nz^{n-1}$$

(3) Let w = log z be defined as log 4 + $(\theta+2k\pi)$ i where k is a fixed integer. Show that except on the cut which is the same as in Ex. (2), the function is analytic and:

$$\frac{d}{dz} \log z = \frac{1}{z}$$

(4) Define

$$\sin z = \frac{e^{zi} - e^{-zi}}{2i}$$

---

*That is to impose the condition that $0 < \theta < 2\pi$.

$$\cos z = \frac{e^{zi}+e^{-zi}}{2}$$

and other trigometrical functions accordingly. Show that they are all periodic functions of period 2π and analytic in the finite z-plane where the function value is finite. All the formulas for the derivatives in elementary calculus can be extended in the present cases.

3. COMPLEX INTEGRATION. CAUCHY'S FUNDAMENTAL THEOREM

All the curves considered in the following sections are Jordan's curves; i.e., continuous curves consisting of a finite number of the arcs defined by the equations:

$$x = \varphi(t) \quad \text{for} \quad a \leqq t \leqq b \quad (\varphi'^2 + \psi'^2 \neq 0)$$

$$y = \psi(t)$$

where $\varphi'(t)$ and $\psi'(t)$ are continuous. A curve is called a <u>simple closed curve</u> if it is closed and meets any straight lines parallel to the axis at most in two points except at extreme values of x and y where a segment may be a part of the boundary such as a square. All the domains in the later discussions are assumed to be bounded by a finite number of nonintersecting simple closed curves.

<u>Definition</u>. For a continuous complex-valued function, the limit:

$$\sum_{j=1}^{n} f(z_j)\Delta z_j$$

(where $z_0 = a$, z_1, ..., $z_n = b$ are a set of points on a curve γ, arranged in the order from the point a to b, and $z = z_j - z_{j-1}$) such that the number n of the subdivisions becomes infinity and the length of each subdivision becomes zero, is denoted by:

$$\int_{\gamma(a,b)} f(z)\, dz$$

Let $f(z) = P(x,y) + iQ(x,y)$ and $z = x + yi$.

Let γ be:

$$x = \begin{cases} x(t) \\ \end{cases} \qquad (a \leqq t \leqq b)$$
$$y = y(t)$$

Then:

$$\int_{\gamma(a,b)} f(z)\,dz = \int_a^b \left(P\frac{dx}{dt} - Q\frac{dy}{dt}\right) dt + i\int_a^b \left(P\frac{dy}{dt} + Q\frac{dx}{dt}\right)dt \tag{16}$$

<u>Theorem 1.</u> <u>(Cauchy's Fundamental Theorem)</u>. If γ is a simple closed curve and $f(z)$ is analytic and single-valued inside the domain D bounded by γ, and if $f(z)$ together with its derivative $f'(z)$ is continuous as $\bar{D} = D+\gamma$, then:

$$\oint_\gamma f(z)\,dz = 0^* \tag{17}$$

The proof of the theorem is almost trivial by (16), Green's theorem and Cauchy-Riemann differential equations:

$$\oint_\gamma f(x)\,dz = \oint_\gamma P\,dx - Q\,dy + i\oint_\gamma P\,dy + Q\,dx$$

$$= -\iint_{\bar{D}} (Q_x + P_y)\,dZ + i\iint_{\bar{D}} (P_x - Q_y)\,dA$$

$$= 0$$

However, the theorem plays a fundamental role in the development of the theory of the function of a complex variable. The theorem is evidently valid, if D is bounded by a finite number non-intersecting

---

*The notation $\oint_\gamma f(z)\,z$ denotes the 'contour integral' along the curve γ counter-clockwise. We also use the notation:

$$\oint_{\gamma -} f(z)\,dz = -\oint_\gamma f(z)\,dz$$

simple closed curve. This remark applies also to the following theorems.

3.1 Cauchy's Residue Formulas

Theorem. If f(z) is analytic in a domain D bounded by γ and continuous on the boundary, (and if f'(z) is continuous on the closed domain D) then, for a fixed z in D,

$$f(z) = \frac{1}{2\pi i}\oint_{\gamma} \frac{f(\zeta)}{\zeta - z}\, d\zeta \tag{18}$$

The derivatives to any order exist and:

$$\frac{d^n f(z)}{dz^n} = \frac{n!}{2\pi i}\oint_{\gamma} \frac{f(\zeta)}{(\zeta+z)^{n+1}}\, d\zeta \tag{19}$$

$(n = 1,2,3, \dots)$

It is remarked that in the more general form of Cauchy's fundamental theorem the condition on the continuity of the derivative on the domain can be removed. See, for instance, E.T. Copson, Introduction to the Theory of a Complex Variable, pp. 61-65.

Proof of the Theorem. For a fixed point z in the domain, let Γ be the circle $|\zeta - z| = \varepsilon$, which lies inside the domain D. By Cauchy's fundamental theorem,

$$\frac{1}{2\pi i}\oint_{\gamma+\Gamma^-} \frac{f(\zeta)}{\zeta - z}\, d\zeta = 0 \tag{20}$$

since,

$$\frac{f(\zeta)}{\zeta - z}$$

is analytic (and, together with the derivatives, is continuous) on the new, closed domain bounded by γ and Γ. On the other hand:

$$\frac{1}{2\pi i}\oint_{\Gamma}\frac{f(\zeta)d\zeta}{\zeta - z} = \frac{1}{2\pi i}\int_0^{2\pi}\frac{f(\varepsilon e^{\theta i}+z)}{\varepsilon e^{\theta i}}\,\varepsilon e^{\theta i} i d\theta$$

$$= \frac{1}{2\pi}\int_0^{2\pi} f(\varepsilon e^{\theta i} + z)\, d\theta \qquad (21)$$

As $\varepsilon \to 0$, the expression (21) tends to the limit $f(z)$. From (20) it follows:

$$\frac{1}{2\pi i}\oint_{\gamma}\frac{f(\zeta)}{\zeta - z}\, d\zeta = f(z)\ , \qquad (22)$$

because the integral on the left-hand side is independent of ε. Next, consider the function defined by the integral:

$$g(z) = \frac{1}{2\pi i}\oint_{\gamma}\frac{f(\zeta)}{\zeta - z}\, d\zeta \qquad (23)$$

The difference quotient

$$Q = \frac{g(z+\Delta z) - g(z)}{\Delta z} = \frac{1}{2\pi i}\oint_{\gamma} f(\zeta)\ \left\{\frac{1}{\zeta-(z+\Delta z)} - \frac{1}{\zeta - z}\right\}\frac{1}{\Delta z}\, d\zeta$$

$$= \frac{1}{2\pi i}\oint_{\gamma} f(\zeta)\ \left\{\frac{1}{[\zeta-(z+\Delta z)](\zeta - z)}\right\} d\zeta \qquad (24)$$

is well defined in domain D provided that $z+\Delta z$ lies inside D. For a fixed z,

$$Q_0 = \frac{1}{2\pi i}\oint_{\gamma} f(z)\,\frac{1}{(\zeta - z)^2}\, d\zeta$$

is a definite quantity. Comparing Q with this quantity, one has:

$$|Q - Q_0| = \frac{1}{2\pi}\oint_{\gamma} f(\zeta)\,\frac{\Delta z(\zeta)\, d\zeta}{(\zeta - z)^2\ \{\zeta-(z+\Delta z)\}}$$

$$= \frac{1}{2\pi}\,|\Delta z| \cdot \left|\oint_{\gamma}\frac{f(\zeta)\, d\zeta}{(\zeta - z)^2\ \{\zeta-(z+\Delta z)\}}\right| \qquad (25)$$

The integral of the last factor in (25) is less than:

$$\frac{ML}{\frac{1}{2}\delta^3}$$

where:

$|f(\zeta)| \leqq M$ for ζ in the closed curve δ,

since:

$f(\zeta)$ is continuous,

L = the length of the contour γ

and:

δ = the minimum distance from z to the contour

provided that:

$|\Delta z| \leqq \frac{1}{2}\delta$.

Thus, it is bounded. As $\Delta z \to 0$, $\frac{dg(z)}{dz}$ exists and is equal to

$$\frac{1}{2\pi i}\oint_\gamma \frac{f(\zeta)}{(\zeta - z)^2}\, d\zeta$$

Now $g(z) = f(z)$. This completes the proof for $n = 1$. The general case can be proved by induction following the same reasoning.

4. RESIDUE CALCULUS

The Cauchy's residue formulas derived in the preceding section are extremely imporant in evaluating many definite integrals. The essential idea is the following theorem which reduces the problem of the contour integration of an analytic function in D (except a finite number of singular points) to the evaluation of the residues at these points. It should be remarked that if $f(z)$ is analytic everywhere in a neighborhood of $z = a$ (i.e., $|z-a| < \varepsilon$ and $z \neq a$) the value of the integral

$$\frac{1}{2\pi i}\oint_\gamma f(z)\,dz$$

is independent of γ by Cauchy's fundamental theorem.

Definition. Let $f(z)$ be analytic everywhere in a neighborhood of $z = a$. The integral

$$\frac{1}{2\pi i}\oint_\gamma f(z)\,dz$$

is defined as the residue of $f(z)$ at $z = a$. This value is shown to be independent of γ by Cauchy's fundamental theorem.

Theorem. If $f(z)$ is analytic on a domain D bounded by γ except at the points $a_1, a_2, \ldots a_m$ inside D, then:

$$\frac{1}{2\pi i}\oint_\gamma f(z)\,dz = \sum_{j=1}^{m} \text{the residue at } a_j\ .$$

This theorem follows immediately from Cauchy's fundamental theorem. The proof is left to the reader. As an illustration of applying this theorem to evaluate some definite integrals, let us consider the following examples:

Example 1.

$$I = \int_{-\infty}^{+\infty} \frac{dz}{(x^2+1)(x^2+4)}$$

Consider the contour integral:

$$\frac{1}{2\pi i}\oint_\gamma \frac{dz}{(z^2+1)(z^2+4)}$$

where γ is the contour as indicated in the following figure:

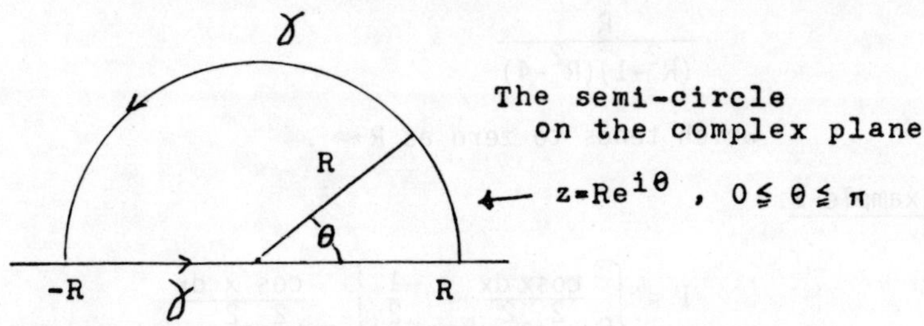

Figure 24

By the above theorem, this contour integral is equal to the residues at $z = i$ and $z = 2i$ which are the only singular points inside the domain bounded by γ. According to Cauchy's theorem, the residue of:

$$\frac{1}{(z^2+1)(z^2+4)} \text{ at } (i)\} \text{ is } \left.\frac{1}{(z+i)(z^2+4)}\right|_{z=i} = \frac{1}{6i}$$

and at 2i is $-\frac{1}{12i}$.

On the other hand, the contour integral can be written as:

$$\frac{1}{2\pi i}\int_0^\pi \frac{Re^{\theta i}\,d\theta}{(R^2e^{2\theta i}+1)(R^2e^{2\theta i}+4)} + \frac{1}{2\pi i}\int_{-R}^{R}\frac{dx}{(x^2+1)(x^2+4)}$$

As $R \to \infty$, the second integral becomes $\frac{1}{2\pi i}$ I and the first integral tends to 0 as it is shown later. Thus:

$$\frac{1}{2\pi i} I = \frac{1}{6i} - \frac{1}{12i} \quad \text{or:} \quad I = \frac{1}{6}\ \pi.$$

The absolute value of the first integral is less than the half of the upper-bound of the absolute value of the integrand for $0 \leqq \theta \leqq \pi$. A simple estimate is done as follows:

$$|R^2e^{2\theta i}+1| \geqq |R^2e^{2\theta i}| - 1 = R^2 - 1\,.$$

Similarly, one can estimate the other factor. Thus, the absolute value of the integrand is, at most, equal to:

$$\frac{R}{(R^2-1)(R^2-4)}$$

which tends to zero as $R \to \infty$.

Example 2.

$$I = \int_0^\infty \frac{\cos x \, dx}{x^2+a^2} = \frac{1}{2} \int_{-\infty}^\infty \frac{\cos x \, dx}{x^2+a^2}$$

where a is a real constant. Consider the contour integral

$$\frac{1}{2\pi i} \oint_\gamma \frac{e^{zi}}{z^2+a^2} \, dz$$

where γ is the same contour as indicated in the above example. It is equal to:

$$\frac{e^{-a}}{2ai}$$

by the theorem as the Cauchy residue formula. On the other hand, the contour integral is:

$$\frac{1}{2\pi} \int_0^\pi \frac{e^{-R \sin \theta + iR \cos \theta} \, Re^{\theta i} \, d\theta}{R^2 e^{2\theta i} + a^2} + \frac{1}{2\pi i} \int_{-R}^R \frac{e^{xi}}{x^2+a^2}$$

The absolute value of the first integral is shown as in the proceding example to be at most equal to:

$$\frac{R}{R^2-a^2}$$

since $e^{-R \sin \theta} \leqq 1$ for $0 \leqq 0 \leqq \pi$.

Thus, it tends to zero as $R \to \infty$. While the second integral becomes $\frac{2}{2\pi i}$ I as $R \to \infty$ since:

$$e^{xi} = \cos x + i \sin x$$

and

$$\int_{-R}^{R} \frac{\sin x}{x^2+a^2}\, dx = 0$$

by the property of an odd function. Consequently,

$$I = \pi i \frac{e^{-a}}{2ai} = \frac{e^{-a}}{2a}$$

Remark. In the evaluating some definite integral such as Ex. 3(b), the estimation of:

$$e^{-R \sin \theta} \leq 1$$

is too rough for the purpose. The following lemma due to Jordan is often very useful for those cases.

Jordan's Lemma

$$\frac{2}{\pi}\theta \leq \sin \theta \leq \theta \quad \text{for} \quad 0 \leq \theta \leq \frac{\pi}{2} .$$

The proof is based on the fact that $\cos \theta$ is a monotonically decreasing function of θ when:

$$0 \leq 0 \leq \frac{\pi}{2}$$

Thus, the average of the function $\cos \theta$ in the interval $(0, \theta_1)$

$$\frac{1}{\theta_1} \int_0^{\theta_1} \cos \theta \, d\theta$$

is a monotonically decreasing function of θ_1. Substituting $\theta_1 = \frac{\pi}{2}$ and taking the limit as $\theta_1 \to 0$, the result follows immediately.

EXERCISES

(1) Find the residue of the functions at the points indicated below:

(a) $\frac{e^{zi}}{z}$ at $z = 0$. Ans. 1

(b) $\dfrac{\sin z}{z^2 - \frac{\pi^2}{4}}$ at $z = \frac{\pi}{2}$ Ans. $\frac{1}{\pi}$

(2) Show that if one defines $f(z) = \dfrac{\sin z}{z}$ for $z \neq 0$ and $f(0) = 1$, $f(z)$ is analytic everywhere.

Find the residue of:

$$\frac{f(z)}{z^3} \quad \text{at} \quad z = 0 . \qquad \text{Ans.} \quad \frac{1}{3!} \left. \frac{d^3 \sin z}{dz^3} \right|_{z=0} = \frac{-1}{3!}$$

Find the value of the following definite integrals by contour integration as indicated:

(a) $\displaystyle\int_0^\infty \frac{\sin x}{x} dx$ by considering $\displaystyle\oint_\gamma \frac{e^{zi}}{z} dz$ where γ is indicated in Fig. 24 with an additional semi-circle around the origin.

Ans. $\frac{\pi}{2}$

(b) $\displaystyle\int_0^\infty \frac{x \sin x}{a^2+x^2} dx$ by considering $\displaystyle\oint_\gamma \frac{ze^{zi}}{a^2+z^2} dz$

Ans. $\frac{\pi}{2} e^{-a}$

(c) Prove that if $c > 0$

$$\frac{1}{2\pi i} \int_{c-i\infty}^{c+i\infty} \frac{a^z}{z^2} dz \quad \begin{cases} \log a & (a > 1) \\ 0 & (0 < a < 1) \end{cases}$$

5. PROPERTIES OF POWER SERIES

The power series plays an important role in theory of analytic functions. The fact that a differentiable function in a domain possesses the derivatives to any order in the domain as shown in Cauchy's residue formula is a rather remarkable result. As a matter of fact, one can even show that it possesses also a Taylor series expansion as shown in the next section.

Consider, now, any power series:

$$\sum_{n=0}^{\infty} a_n z^n \tag{26}$$

The convergence of such a series is defined in the elementary calculus.

<u>Theorem</u>. If the infinite series (26) converges at a complex value $z = z_0 \neq 0$, then the series converges absolutely and uniformly for $|z| \leqq a < |z_0|$ to an analytic function f(z). Furthermore, the series obtained by differentiating (26) term by term has the same properties and is equal to $\frac{df(z)}{dz}$.

<u>Proof</u>. Let

$$\left|\frac{a}{z_0}\right| = r \ (<1) \ .$$

Since the series (26) is convergent at $z = z_0$, it follows that:

$$|a_n z_0^n| \to 0 \quad \text{as} \quad n \to \infty \ .$$

Thus,

$$|a_n z_0^n| < M \quad \text{for all} \quad n,$$

where M is a constant. Now, the series:

$$\sum_{n=1}^{\infty} r^n \quad \text{converges.}$$

Thus, for any given ε,

$$\sum_{n=N}^{N+p} r^n < \varepsilon, \quad (p > 0)$$

provided N is large enough. Therefore,

$$\sum_{n=N}^{N+p} |a_n z^n| = \sum_{n=N}^{N+p} |a_n z_0^n| \left|\frac{z}{z_0}\right|^n < M \sum_{n=N}^{N+p} r^n < M\varepsilon \ .$$

By Cauchy's necessary and sufficient condition, the series converges uniformly and absolutely for

$$|z| \leqq a < |z_0|.$$

The rest of the theorem follows from the following facts:

(i) If f(z) is analytic function in D,

$$\int_0^z f(z_1)\, dz_1$$

defines a function of z independent of the path since the contour integration of $f(z_1)$ along any closed contour is zero by Cauchy's fundamental theorem. On the other hand, the derivative of the integral is equal to the integrand as it is readily verified from the definition and the condition that $f(z_1)$ is continuous.

(ii) Any uniformly convergent series can be integrated term by term. Now, for

$$|z| \leq a < |z_0|,$$

let

$$F(z) = \sum_{n=0}^{\infty} a_n z^n$$

The fact that the series:

$$\sum_{n=1}^{\infty} n a_n z^n$$

converges uniformly and absolutely follows from the same argument as in the above proof. Let:

$$G(f) = \sum_{n=0}^{\infty} n a_n z^{n-1}$$

Now, by (ii)

$$\int_0^z G(z) = \sum_{n=1}^{\infty} a_n z^n = -a_0 + F(z)$$

By (i), it follows that F(z) is analytic and:

$$\frac{dF(z)}{dz} = G(z) = \sum_{n=1}^{\infty} na_n z^{n-1}$$

This completes the proof.

It is almost trivial to add that F(z) is continuous function of z. The series:

$$\sum_{n=1}^{\infty} a_n |z|^n$$

can be made as small as one pleases if $|z|$ is small enough.

For any infinite series (26), there is an R, defined as <u>the radius of convergence</u> of the series such that the series (26) converges for $|z| < R$ and diverges for $|z| > R$. Little can be said about the convergence at a point on $|z| = R$. Of course, R can be 0, a finite positive quantity or $+\infty$. The circle $|z| = R$ is called the circle of convergence. <u>Within the circle of convergence, the function defined by the infinite series is always analytic.</u>

6. LAURENT'S SERIES

Probably the most striking fact is that there are only three types of an isolated singular point of an analytic function. This is a consequence of the following:

<u>Theorem</u>. If f(z) is analytic everywhere in the neighborhood of z=a except possibly at z = a, then in the neighborhood, the function possesses an expansion in terms of the positive and the negative powers of z - a, namely:

$$f(z) = \sum_{n=-\infty}^{\infty} A_n (z-a)^n$$

where:

$$A_n = \frac{1}{2\pi i} \oint_\gamma \frac{f(z)}{(z-a)^{n+1}} dz \qquad (n = 0, \pm 1, \pm 2, \pm 3, \ldots) \tag{27}$$

and γ is any simple closed contour around a.

Proof. For any fixed point z in the neighborhood $|z-a| < R$ and $z \neq a$, let γ be the circle $|\zeta - a| = R$ and γ_1 be the circle $|\zeta - a| = r$ ($< |z-a|$). By Cauchy's residue formula, one has:

$$f(z) = \frac{1}{2\pi i} \oint_\gamma \frac{f(\zeta)}{\zeta - z} d\zeta - \frac{1}{2\pi i} \oint_{\gamma_1} \frac{f(\zeta)}{\zeta - z} d\zeta \tag{28}$$

where both of the contour integrals are taken along the circles counter-clockwise. For:

$|\zeta - a| = R$ $\left|\frac{z-a}{\zeta-a}\right| = \frac{|z-a|}{R} < 1$, one sees that:

$$\frac{1}{\zeta - z} = \frac{1}{\zeta - a} \cdot \frac{1}{1 - \frac{z-a}{\zeta-a}} \sum_{n=0}^{\infty} \frac{1}{(\zeta-a)^{n+1}} (z-a)^n \tag{29}$$

The series (29) converges uniformly for all values of ζ on $|\zeta - a| = R$. Multiplying the series (29) by $f(\zeta)$ and integrating term by term, one concludes that the first integral in (28) is:

$$\sum_{n=0}^{\infty} A_n (z-a)^n$$

where A_n is given by (27).

A similar argument shows that the second integral in (28) is:

$$\sum_{m=1}^{\infty} \frac{A'm}{(z-a)^m} \tag{30}$$

where:

$$A'_m = \frac{1}{2\pi i} \oint_{\gamma_1} (\zeta-a)^{m-1} f(\zeta) d\zeta , \qquad (m = 1, 2, \ldots)$$

Now set m = -n and set:

$$A_n = A'_m = \frac{1}{2\pi i}\oint_{\gamma_1} \frac{\zeta}{(\zeta-a)^{n+1}}\,d\zeta \qquad (n = -1, -2, \ldots)$$

By Cauchy's fundamental theorem:

$$A_n = \frac{1}{2\pi i}\oint_{\gamma} \frac{f(\zeta)}{(\zeta-a)^{n+1}}\,d\zeta$$

This completes the proof.

Corollary. (Taylor's Series) If, in addition, $f(z)$ is analytic at $z = a$, then $A_n = 0$ for $n = -1, -2, \ldots$ and

$$f(z) = \sum_{n=0}^{\infty} \frac{f^{(n)}(a)}{n!}(z-a)^n \tag{31}$$

By (27), this is direct consequence of Cauchy's fundamental theorem and Cauchy's residue formulas. It should be remarked that the Taylor series converges at least inside the circle with the center at $z = a$, which touches the boundary of the domain of analyticity of $f(z)$. In view of the formula (27), the following theorem is evident. However, it is of paramount in the theory of the residue calculus.

Theorem. The residue of $f(z)$ at $z = a$ is A_{-1} is the coefficient of $(z-a)^{-1}$ in the Laurent's series.

6.1 The Types of Isolated Singularities

As far as the number of the negative powers in the Laurent's series is concerned, there are only three possibilities:

(i) No Negative Powers

This type of singularity is characterized by the property that the limit $f(z)$, as $z \to a$, exists. For instance,

$$f(z) = \sin z/z \qquad (z \neq 0) .$$

The function is not defined at z = 0. If, now, f(0) is defined as

$$\lim_{z\to 0} \frac{\sin z}{z} = \lim_{z\to 0} \sum_{\substack{m=1 \\ (n=2\,m-1)}}^{\infty} (-1)^{m-1} \frac{z^{n-1}}{n!} = 1 ,$$

it is seen that f(z) is also analytic at z = 0. For such a reason, the singularity is called a <u>removable singularity</u>.

(ii) <u>A Finite Number of Negative Powers.</u>

It is characterized by the property that:

$$f(z) = \left| \frac{1}{(z-a)^m} \sum_{n=0}^{\infty} a_n (z-a)^n \right| \to \infty$$

as $z \to a$, where m is the number of the highest negative powers. <u>It is called a pole</u>. The number m is defined as the order of the pole. According to m = 1, 2, ..., the singularity is defined as a simple pole, a double pole, etc. For instance:

$$f(z) = \frac{\sin z}{z^3}$$

has a double pole at z = 0.

(iii) <u>Infinitely Many Terms of the Negative Powers.</u>

The behavior of this type of singularity is discovered by <u>Weierstrass</u>. For any preassigned complex quantity k and any preassigned positive number ε, there is always a value z_1 of z(≠a) in the neighborhood of z = a, however small it may be, such that:

$$|f(z) - k| < \varepsilon$$

If a is the limiting point of the zeros of f(z) - k, the theorem is evident. Now, assume that:

$$g(z) = \frac{1}{f(z)-k}$$

is analytic for $0 < |z-a| \leq r$ where r is a positive number (small enough). It is to show that there is a point z_1 in this neighborhood, such that:

$$|g(z_1)| > \frac{1}{\varepsilon}$$

Suppose, on the contrary, $|g(z)| \leq \frac{1}{\varepsilon}$. Consider:

$$\sum_{n=-1}^{-\infty} B_n (z-a)^n$$

where:

$$B_n = \frac{1}{2\pi i} \oint_{|z-a|=\rho} \frac{g(z)}{(z-a)^{n+1}} dz \qquad (32)$$

and $0 < \rho < r$. The expression (32) gives immediately the inequality

$$|B_n| \leq \frac{\rho^{-n}}{\varepsilon} \; (n = -1, -2, \ldots,)$$

Now, B_n is the coefficient of a negative power in the Laurent's series and is independent of ρ. On the other hand, as $\rho \to 0$, $B_n \to 0$. Thus, $g(z)$ is analytic at $z = a$. This would mean that:

$$f(z) = k + \frac{1}{g(z)}$$

is either analytic or has a pole at $z = a$. Because an analytic function ($\not\equiv 0$) has, at most, a zero of finite order at a point (as it is easily seen from the power series). Therefore, it is impossible. This completes the proof.

Such a singularity is called an <u>essential singularity</u>. For instance $f(z) = e^{1/z}$ has an essential singularity at $z = 0$. <u>The singularity of $f(z)$ at infinity is defined by the singularity of $F(\zeta) = f(1/\zeta)$ at $\zeta = 0$.</u>

6.2 L'Hospital Rule

The L'Hospital rule for the functions of a real variable in the elementary calculus can be generalized to functions of a complex variable, which is analytic (or has a pole) at the point. The proof is left to the reader. The rule is a direct consequence of Taylor's series at the point. In most cases the limit of a quotient is also easily found by Laurent's series. For instance,

$$\lim_{z\to 0} \frac{z^2}{\sin^2 z} = \lim_{z\to 0} \frac{1^1}{(1 - \frac{z^2}{3!} + \frac{z^4}{5!} + \ldots)^2} = 1$$

In view of the discussion in Section 6.1, the function $\frac{z^2}{\sin^2 z}$ has a removable singularity at $z = 0$.

6.3 Applications to the Residue Calculus

The second theorem in Section 6.1 is particularly useful in the residue calculus. In the case of a simple pole, one derives immediately:

$$\text{(The residence at a)} = A_{-1} = \lim_{z\to a} f(z)\,(z-a) \qquad (33)$$

For instance:

$$\text{(The residence of } \frac{z}{\sin^2 z} \text{ at 0)} = \lim_{z\to 0} \frac{z^2}{\sin^2 z} = 1 \quad .$$

In the case of the pole of higher order, the computation for the residue, very often, is rather complicated by directly applying Cauchy's residue formulas. It is then more advisable to use Laurent's expansion. As far as the residue is concerned, the only coefficient needed is A_{-1}. For instance, consider the function:

$$\frac{1}{\sin^3 z} \quad \text{at} \quad z = 0 \; .$$

Now,

$$\frac{1}{z^3\left(1 - \frac{z^2}{3!} + \frac{z^4}{5!} + \ldots\right)^3} = \left\{1 - \left(\frac{z^2}{3!} - \frac{z^4}{5!} + \ldots\right)\right\}^{-3} \cdot \frac{1}{z^3}$$

and:

$$\left\{1 - \left(\frac{z^2}{3!} + \ldots\right)\right\}^{-3} = 1 + 3\,\frac{z^2}{3!} + \ldots$$

Therefore, (the residue of $\frac{1}{\sin^3 z}$ at 0) = $\frac{1}{2}$.

Observe that one would have to compute $\frac{d^2}{dz^2} \frac{z^3}{\sin^3 z}$ at $z = 0$, if he applies the Cauchy residue formula.

EXERCISES

(1) From the formula:

$$\frac{1}{2\pi i} \oint_C \frac{(1+z)^n}{z^{(k+1)}}\, dz = \frac{n(n-1)\ \ldots\ (n-k+1)}{k!}$$

where n and k are positive integers and C is a circle around the origin, deduce the relation:

$$\int_0^{\pi} (2 \cos u)^{n+k} \cos (n-k)\, u\, du = \pi \frac{(n+!)(n+2)\ldots(n+k)}{k!}$$

and:

$$\int_{-1}^{+1} \frac{x^{2n}}{(1-x^2)^{\frac{1}{2}}}\, dx = \pi\, \frac{1\cdot 3\cdot 5\ \ldots\ (2n-1)}{2\cdot 4\cdot 6 \ldots\ 2n}$$

(2) Evaluate the following definite integral

$$\int_0^{\infty} \frac{x \log x}{(1+x^2)^3}\, dx$$

(3) Show that:

$$\int_0^{\infty} \frac{\log^2 z}{1+z^2}\, dz = \frac{\pi^3}{8}$$

7. CAUCHY'S EXPANSION THEOREM

Definition 1. If a function is analytic everywhere in the finite plane except, possibly, poles, it is called a meromorphic function.

Meromorphic functions occur very often on many occasions. To find the inverse transform of the solution of a boundary value problem, the inverse transform is immediately obtained if it is a rational function. Thus, the Cauchy's expansion theorem which enables one to express a meromorphic function in a series of rational functions is particularly useful.

Definition 2. The part of all negative powers

$$\sum_{n=1}^{\infty} A_{-n}/(z-a)^n$$

of the Laurent's series of a function f(z) is defined as the principal part of f(z) at z = a. It is denoted by $P_f(a)$.

Theorem. Let $C_1, \ldots, C_n$ be a family of concentric circles* with radii R_n such that $R_n \to \infty$ as $n \to \infty$ and none of them passes the poles of a function f(z). If:

(i) f(z) is a meromorphic function with the poles at $a_1, a_2, \ldots, a_n \ldots$;

---

*This family may be replaced by any family of simple closed curves C_n such that the length of C_n is of the order R_n where R_n is the minimum distance from a fixed point, say the origin, to the curve C_n.

A more general result is immediately obtained from the proof of the theorem that the condition (iii) is replaced by the following:

(iia) $f(R_n e^{\theta i}) \to 0$

uniformly in any closed portion of the θ-interval [0, 2π] which does not include one of the finite exceptional values of θ.

(iii) $|f(z)|$ is uniformly bounded on C_n for all n.

and

(ii) $\lim_{R_n \to \infty} |f(R_n e^{\theta i})| = 0$ uniformly with respect to θ in $0 \leq \theta \leq 2\pi$.

then

$$f(z) = \sum_m P_f(a_m) \tag{34}$$

where $P_f(a_m)$ is the principal part of f(z) at $z = a_m$.

Proof. Let a_1, ... a_m be all the poles of f(z) inside the circle C_n. There are only a finite number of poles inside C_n since all the singularities of f(z) are poles and a pole is an isolated singularity. Otherwise by Weierstrass' theorem there would be a limiting point of the poles and the limiting point would be a singular point which is not isolated. Now, let γ_1, ... γ_m be the small circles with the centers respectively at a_1, ... a_m. By Cauchy's residue formula, for any regular point z inside it follows:

$$f(z) = \frac{1}{2\pi i} \int_{C_n + \sum_{j=1}^{n} \gamma_j} \frac{f(\zeta)}{\zeta - z} d\zeta \tag{35}$$

On:

$$\gamma_j \, , \; |\zeta - a_j| < |z - a_j|$$

and:

$$\frac{1}{\zeta - z} = \frac{-1}{z - a_j} \frac{1}{1 - \frac{\zeta - a_j}{z - a_j}} = - \sum_{k=0}^{\infty} \frac{(\zeta - a_j)^k}{(z - a_j)^{k+1}}$$

Therefore:

$$\frac{1}{2\pi i} \oint_{\gamma_j} \frac{f(\zeta) d\zeta}{\zeta - z} = - \sum_{k=1}^{\infty} \frac{A_{-k}}{(z - a_j)^k} = P_f(a_j)$$

where:

$$A_{-k} = \frac{1}{2\pi i} \oint_{\gamma_j} \frac{f(\zeta)\, d\zeta}{(\zeta - a_j)^{k+1}}$$

On the other hand:

$$\left| \frac{1}{2\pi i} \oint_{C_n} \frac{f(\zeta)}{\zeta - z}\, d\zeta \right| \leq \left\{ \begin{matrix} \max \ |f(\zeta)| \\ |\zeta| = R_n \end{matrix} \right\} \frac{2\pi R_n}{R_n - |z|}$$

which tends to zero as $n \to \infty$. Now let n tend to infinity, the equation (35) gives immediately the result (34). This completes the proof.

It should be remarked that Cauchy's expansion theorem has been very much extended. For further references, see, for instance, Titchmarsh, Theory of Functions, pp. 110-112. One special case which is of worthy notice is that instead of the condition (ii), the following condition is satisfied:

$$\frac{|f(Re^{\theta i})|}{R^p} \to 0$$

uniformly with respect to θ in $0 \leqq \theta \leqq 2\pi$ for some positive integer p. Then one expands:

$$\frac{1}{\zeta - z} \quad \text{as} \quad \sum_{k=1}^{p} \frac{z^{k-1}}{\zeta^k} + \frac{z^p}{\zeta^p(\zeta - z)}$$

and the contour integration on C_n

$$\frac{1}{2\pi i} \oint_{C_n} \frac{f(\zeta)}{\zeta -}\, d\zeta = \sum_{k=1}^{p} b_k z^{k-1} + z^p \oint_{C_n} \frac{f(\zeta) d\zeta}{\zeta^p(\zeta - z)} \qquad (36)$$

The last integral tends to zero by the assumption as $n \to \infty$. The value of b_k (k-1, ... p-1) can be evaluated by the residue theory from the integral:

$$b_k = \frac{1}{2\pi i}\oint_{C_n} \frac{f(\zeta)d\zeta}{\zeta^k} \quad (k = 1, \ldots p)$$

Example. $f(z) = \tan z$. The poles are at $z = \frac{\pi}{2} + m\pi$ ($m = 0, +1, +2, \ldots$). Let C_n be the squares with vertices at $n\pi(1 \pm i)$ ($n = 1, 2, \ldots$). On the side from $-n\pi + in\pi$ to $n\pi + in\pi$

$$|\tan z| = \left|\frac{e^{x+in\pi} - e^{-x-in\pi}}{e^{x+in\pi} + e^{-x-in\pi}}\right| \leqq \left|\frac{e^{x} - e^{-x}}{e^{x} + e^{-x}}\right|$$

which is bounded (independent of n) for $-n\pi \leqq x \leqq n\pi$. Similarly it is uniformly bounded on the side from $-n\pi-in\pi$ to $n\pi-in\pi$. On the side $n\pi-in\pi$ to $n\pi+in\pi$,

$$|\tan z| = \left|\frac{e^{n\pi+iy} - e^{-n\pi-iy}}{e^{n\pi+iy} + e^{-n\pi-iy}}\right|$$

$$\leqq \frac{e^{n\pi} + e^{-n\pi}}{e^{n\pi} - e^{-n\pi}} \leqq \frac{1 + e^{-2n\pi}}{1 - e^{-2n\pi}} \leqq \frac{1 + e^{-2\pi}}{1 - 3^{-2\pi}}$$

Similarly, it is bounded uniformly for the side from $-n\pi - in\pi$ to $-n\pi + in\pi$. Thus:

$$\left|\frac{\tan z}{z}\right| \to 0 \text{ on } C_n \text{ uniformly as } n \to \infty.$$

This is the case $p = 1$. Now, by the residue theory:

$$b_1 = \frac{1}{2\pi i}\oint_{C_n} \frac{f(\zeta)d\zeta}{\zeta} = \sum_{m=-n+1}^{n-1} \text{residues at } \left(m\pi + \frac{\pi}{2}\right) + f(0)$$

$$= \sum_{m=-n+1}^{n-1} \lim_{\zeta \to (m\pi + \frac{\pi}{2})} \frac{\zeta - \frac{2m+1}{2}\pi}{\zeta} \tan \zeta$$

$$= \sum_{m=-n+1}^{n-1} \frac{-1}{\frac{2m+1}{2}\pi}$$

One readiliy shows again that

$$P_{\tan z}\left(\frac{2m+1}{2}\pi\right) = (-1)/z - \frac{2m+1}{2}\pi)$$

Therefore one obtains:

$$\tan z = \sum_{m=-\infty}^{\infty} \left\{\frac{-z}{z - \frac{2m+1}{2}} + \frac{1}{\frac{2m+1}{2}\pi}\right\}$$

$$= \sum_{m=0}^{\infty} \frac{-2z}{z^2 - \left(\frac{2m+1}{2}\right)^2 \pi^2} \tag{37}$$

EXERCISES

Obtain the expansions for the following functions:

(1) $\sec z = 2\pi \sum_{n=0}^{\infty} \frac{(-1)^n (n+1/2)}{(n+1/2)^2 \pi^2 - z^2}$

(2) $\operatorname{cosec}^2 z = \sum_{n=-\infty}^{\infty} \frac{1}{(z-n\pi)^2}$

(3) $1/(\cos az - \cos bz)$, $a \neq \pm b$

(4) $1/(1-e^z)$

(5) $\sinh (a z) \sinh z$, $0 < a < 1$

(6) Let a = integer. Establish the identity

$$\sum_{n=-\infty}^{\infty} \frac{1}{(a+n)^2} = \pi^2 \operatorname{cosec}^2 a$$

by considering the contour integration of the function

$\cot \pi z/(z+a)^2$.

(7) Sum the series

(a) $\sum_{n=1}^{\infty} a/(n^4+a^4)$

(b) $\sum_{n=1}^{\infty} n^2/(n^4+a^4)$

(8) Prove that

$$\sum_{n=1}^{\infty} 1/n^3 \sin (n\pi\sqrt{2}) = -13\pi^3/\sqrt{2}\cdot 360$$

HINT. Consider the integral:

$$\frac{1}{2\pi i}\oint \frac{1}{z^3}\frac{dz}{\sin \pi z \sin \theta\pi z}$$

where $\theta = \sqrt{2}^{-1}$

8. THE EXPANSION THEOREM OF AN ARBITRARY FUNCTION IN TERMS OF EIGENFUNCTIONS

We are now ready to prove the theorem in Section 4, Chapter II. That all the eigenvalues are real has been shown in Chapter II. Consider the following transformation of the variables:

$$z = \frac{1}{k}\int_a^x \left\{\frac{r(x)}{p(x)}\right\}^{\frac{1}{2}} dx \quad \text{and} \quad u = \{p(x)\ r(x)\}^{\frac{1}{4}}\ y \tag{38}$$

where

$$k = \int_a^b \left\{\frac{r(x)}{p(x)}\right\}^{\frac{1}{2}} dx \ > \ 0 .$$

The equation (27) on p. , Chapter II, becomes:

$$\frac{d^2u}{dz^2} + \{\rho^2 - g(z)\}\ u = 0 \tag{39}$$

where:

$$\rho^2 = k^2\lambda\ ,\quad g(z) = \frac{\frac{d^2}{dx^2}(pr)^{\frac{1}{4}}}{(Pr)^{\frac{1}{4}}} + k^2\ \frac{q(x)}{r(x)}$$

The conditions (28) are of the same form:

$$\alpha u(0) + \beta u'(0) = 0$$

$$\gamma u(1) + \delta u'(1) = 0 \tag{40}$$

such that

$$(\alpha,\beta) \neq (0,0) \quad \text{and} \quad (\gamma,\delta) \neq (0,0) \quad .$$

The equation (39) can be rewritten in the form:

$$\frac{d^2u}{dz^2} + \rho^2 u = g(z)\, u \tag{41}$$

and its general solution is, by the variation of parameters, given by:

$$u(z) = A \cos \rho z + B \sin \rho z + \frac{1}{\rho}\int_0^z \sin \rho(z-t) g(t) u(t)\, dt \tag{42}$$

To be more definite, the solution which satisfies the conditions

$$u(0) = \beta, \quad u'(0) = -\alpha \quad ,$$

is:

$$u(z) = \beta \cos \rho z + \frac{\alpha}{\rho} \sin \rho z + \frac{1}{\rho}\int_0^z \sin \rho(z-t) g(t) u(t)\, dt \tag{43}$$

The fundamental existence theorem in the theory of ordinary equations* affirms that u(t) is bounded in absolute value in the entire interval. The solution is unique and for a fixed value z, it is analytic in ρ for all finite complex values of ρ(i.e., an entire function of ρ). Let $M(\rho)$ be the least upper-bound of u for $0<z\leqq 1$, then for all positive values of ρ,

$$|u(z)| < \beta + \frac{|\alpha|}{\rho} + \frac{M(\rho)}{\rho}\int_0^1 |g(t)|\ dt \ .$$

Since u(z) is continuous in the closed interval $0 \leqq z \leqq 1$, it attains its maximum, thus:

---

*See, for instance, Ince, Ordinary Differential Equation, pp. 73-81.

$$M(\rho) \leqq (|\beta| + \frac{|\alpha|}{\rho}) \{1 - \frac{1}{\rho}\int_0^1 |g(t)| \ dt\} \tag{44}$$

provided that ρ is large enough. This result will be used later.

Now this particular solution satisfies the first boundary condition of (40) automatically. If the second condition is applied, it is found that the eigenvalues are then the zeros of an entire function of ρ. Because an analytic function has only isolated zeros in the domain of analyticity, the entire function is either identically zero or it has a discrete set of zeros, which has, at the most, one limiting point* at infinity.

The second boundary condition of (40) can be explicitly written as:

$$f(\rho) = (\frac{\alpha\gamma}{\rho^2} + \delta\beta) \sin\rho + (\alpha\delta-\gamma\beta) \frac{1}{\rho} \cos\rho$$

$$- \frac{1}{\rho}\int_0^1 g(t)u(t) \ [\frac{\gamma}{\rho} \sin \rho(1-t) + \delta \cos \rho(1-t)] \ dt = 0 \tag{45}$$

One has to discuss this equation in the following cases:

(i) If $\delta\beta \neq 0$, it is seen that $f(\rho)$ is alternatively positive and negative when $\rho = (n+\frac{1}{2})\pi$ for large integral values n. Thus $f(\rho)$ does have a discrete set of zeros.

---

*A point z_0 is defined as a limiting point of the set $z_n (n = 1,2,...)$ if, in any neighborhood of z_0, there is always a point belonging to the set other than z_0 itself.

**In general, a complex-valued function $F(\rho)$ is defined as $O(|\rho|^m)$ if there exists a constant C such that $F(\rho) \leqq C|\rho|^m$ for large values of $|\rho|$ That the eigenvalues λ are bounded below is a consequence of the facts: (i) Any two eigenfunctions belonging to different eigenvalues are orthogonal; (ii) For a negative value λ with absolute value large enough, the solution satisfying the conditions [$u(0) = \beta$, $u'(0) = -\alpha$] has no zeros in [0,1]. The latter (ii) follows from the lemma 1 on p.

(ii) If $\delta = 0$ and $\beta \neq 0$, (or $\delta \neq 0$ and $\beta = 0$), the cosine term becomes predominant (since $\gamma \neq 0$) for large values of ρ. The same result follows.

(iii) If $\delta = 0$ and $\beta = 0$, the inequality (44) gives that the integral in (45) is $O(1/\rho^2)$**. Now, $\alpha\gamma \neq 0$. The sine term is again predominant. So the same result follows.

This completes the proof of the existence of a discrete set of infinitely many eigenvalues, at the most, a finite number of which are negative.

8.1 Asymptotic Order of Eigenvalues and Zeros of Eigenfunctions

A more precise assertion can be made concerning the asymptotic expression of the nth eigenvalue $\lambda_n = (\frac{\rho_n}{k})^2$ as a function of n. The whole idea is based on the following:

<u>Auxiliary Theorem</u>. If the eigenvalues of the Sturm-Liouville system (39) and (40) are arranged in the increasing order:

$$\lambda_0, \lambda_1, \ldots\ldots\ldots\ldots, \lambda_m \ldots\ldots\ldots\ldots$$

and if the corresponding eigenfunctions are:

$$u_0(z), u_1(z), \ldots\ldots\ldots\ldots, u_m(z) \ldots\ldots\ldots\ldots$$

then $u_m(z)$ has exactly m zeros in the interval $0 < z < 1$.

The main results used to prove this theorem are the following two lemmas:

<u>Lemma 1</u>. The number of zeros in $0 < z \leq 1$ of a solution satisfying the initial conditions $u(0) = \beta$ and $u'(0) = -\alpha$, $(\alpha, \beta) \neq (0,0)$, increases with the function $G(z) = \lambda - g(z)$.

---

* and **. The footnote is given on the preceding page.

Proof. First of all, the number of zeros of any solution u(z) of (39), which is not identical zero, is always finite. Otherwise, there is a sequence of zeros of u(z):

$$z_1, z_2, \ldots\ldots\ldots\ldots z_n, \ldots\ldots\ldots\ldots,$$

which has a limiting point at "a" in [0,1]. Then by the law of the mean in calculus,

$$u(z_n) = u(a) + u'(a + \theta_n(z_n - a))$$

where $0 < \theta < 1$. Since $u(a) = u(z_n) = 0$ and $u'(z)$ is a continuous function of z, one concludes immediately that $u'(a) = 0$. By the fundamental thoerem of existence and uniqueness, $u(z) \equiv 0$. This is a contradiction.

Let $v_s(z)$ $(s = 1,2)$ be the solutions satisfying the initial conditions corresponding to $G(z) = G_s(z) = \lambda_s - g_s(z)$. Assume that $G_1 \geqq G_2$ for $0 \leqq z \leqq 1$ and $G_1 \not\equiv G_2$. Let

$$z_1 < z_2 < \ldots < z_m$$

be the zeros of

$$v_2(z) \quad \text{in} \quad 0 < z \leqq 1 \ .$$

From the equations for $v_1(z)$ and $v_2(z)$, it follows readily:

$$v_1(z)\, v_2'(z) - v_2(z) v_1'(z) \Big|_{z_i}^{z_{i+1}} = \int_{z_i}^{z_{i+1}} (G_1 - G_2)\, v_1 v_2 \, dz \tag{46}$$

Now, $v(z) \neq 0$ in $z_i < z < z_{i+1}$. One may assume, without loss of generality, $v_2(z) > 0$ in the open interval.

Suppose that $v_1(z) \neq 0$ in the same interval. Let $v_1(z) > 0$ in the open interval. The right-hand side of (46) is positive. Since $v_s(z_i) \geqq 0$, $v_s(z_{i+1}) \geqq 0$, $v'(z_i) \leqq 0$ and $v_s(z_{i+1}) \geqq 0$, the left-hand side of (46) would be $\leqq 0$. This contradiction leads to the fact that $v_1(z)$ has at least one zero in $z_i < z < z_{i+1}$.

If $z = 0$ is also a zero of $v_2(z)$, the lemma is then trivial. Otherwise, one can prove that between 0 and z_1, there is at least one zero of $v_1(z)$. This is done (by Picone) as follows:

$$\frac{d}{dz}\left[\frac{v_2}{v_1}(v_1v_2' - v_2v_1')\right] = \frac{v_2}{v_1}(G_1-G_2)\, vv + \frac{(v_1v_2'-v_2v_1')^2}{v_1^2} . \quad (47)$$

If $v_1(z) > 0$ in $0 < z < z_1$, one obtains:

$$v_2^2\left(\frac{v_2'}{v_2} - \frac{v_1'}{v_1}\right)\Big|_0^{z_1} = \int_0^{z_1}(G_1-G_2)\, v_2^2\, dz + \int_0^{z_1}\frac{(v_1v_2'-v_2v_1')^2}{v_1^2}\, dz. \quad (48)$$

The right-hand side of (48) is positive. Now, as:

$$v_1(0) = v_2(0) = \beta \quad \text{and} \quad v_1'(0) = v_2'(0) ,$$

The argument is applied to the limiting case when the upper limit approaches z_1. This contradiction shows that $v_1(z)$ must have at least a zero in $0 < z < z_1$.
Thus, the proof of the lemma 1 is completed.

<u>Lemma 2</u>. Let c be such that $v_1(c)v_2(c) \neq 0$. If both $v_1(z)$ and $v_2(z)$ have the same number of zeros in $a < z < c$, then:

$$\frac{v_1'(c)}{v_1(c)} < \frac{v_2'(c)}{v_2(c)}$$

(Here $v_1(z)$ and $v_2(z)$ are the solutions in the proof of the lemma 1).

<u>Proof</u>. Let z_i be the zero next beore c of $v_2(z)$. Then $v_1(z)$, by the lemma 1, must have at least i zeros. While the number of zeros of $v_1(z)$ is the same as that of $v_2(z)$ in the whole interval, it must have exactly i zeros between a and z_j. Therefore, $v_1(z_j) \neq 0$. Thus, the Picone's formula (47) gives immediately the result (49). This proves the lemma 2.

Proof of Auxiliary Theorem. For sufficiently small value of λ which is negative such $\leq -m^2$, the solution $u = u(z,\lambda)$ of (39) satisfying the initial conditions $u(0) = \beta$, $u'(0) = -\alpha$ has no more zeros in $0 < z \leq 1$ than the solution $y(z)$ of the equation:

$$y'' - m^2 y = 0$$

with the same initial conditions. Now $y(z)$ can be found explicitly:

$$y(z) = \beta \cos h\ mz - \frac{\alpha}{m} \sin h \qquad (50)$$

If $\beta = 0$, it is clear that $y(z)$ has no zeros in $0 < z \leq 1$. If $\beta \neq 0$, one can take m so large that $\cot h\ (mz) > \frac{1}{2}$ for $0 < z \leq 1$ and:

$$\left|\frac{\alpha}{m}\right| < \frac{1}{2}$$

So, $y(z)$ has again no zero in $0 < z < 1$. Thus, for $|\lambda|$ large enough and $\lambda < 0$, $u(z,\lambda)$ has no zeros in $0 < z \leq 1$.
Now as λ increases, the number of zeros of $u(z,\lambda)$ in $0 < z \leq 1$ increases and each zero tends to move in the direction of the end-point 0. Consequently, for a certain value of λ, say $\lambda = \mu_1$, the solution acquires a first additional zero which appears at the end-point 1* and then travels to the left, as λ increases, towards 0. For the value $\lambda = \mu_2$, another zero appears and so on. Thus, there exists a sequence of numbers:

$$\mu_1,\ \mu_2,\ \ldots\ldots\ldots,\ \mu_m\ \ldots$$

*If the zero appears at some point z_1 in $0 < z < 1$ then $\partial/\partial z_1\ u(z_1, \mu_1) \neq 0$, and $u(z_1+\varepsilon,\ \mu_1)\ u(z_1-\varepsilon,\ \mu_1) < 0$. Since $u(z,\lambda)$ is continuous in λ, $u(z_1+\varepsilon,\lambda)\ u(z_1-\varepsilon,\lambda) < 0$ for all value of λ close enough to μ_1. This would mean a contradiction that $u(z,\lambda)$ has a zero in the concerned interval for some $\lambda < \mu_1$.

Such that when

$$\mu_m < \lambda < \mu_{m+1} ,$$

the equation (39) with the above mentioned initial conditions admits of a unique solution which has exactly m zeros in $0<z<1$. This solution satisfies the first boundary condition of (40) automatically. The second boundary condition is:

$$\gamma u(1,\lambda) + \delta u'(1,\lambda) = 0 .$$

If $\delta = 0$, then $\lambda_{m-1} = \mu_m$ and the theorem is trivial. Assume, now, that $\delta \neq 0$. From the lemma 2,

$$\frac{u'(1,\lambda)}{u(1,\lambda)}$$

is monotonically decreasing function of λ from $+\infty$ to $-\infty$ in $\mu_{m-1}< \lambda < \mu_m$. Since $u(1, \mu_{m-1}) = 0$, consequently $u'(1, \mu_{m-1}) \neq 0$. Thus, the second boundary condition will be satisfied for one and only one value λ_{m-1} in the λ-interval $\mu_{m-1} < \lambda < \mu_m$. Thus $u_m(z)=u(z,\mu_m)$ has m zeros in $0 < z < 1$. This proves the auxiliary theorem.*

8.1.1 The asymptotic expression of the eigenvalues. Because of (44), one can write:

$$u(t) = \beta \cos \rho t - \frac{\alpha}{\rho} \sin \rho t + \frac{\beta\sigma(t,\rho)}{\rho} + \frac{\nu(t,\rho)}{\rho} \tag{51}$$

where $\sigma(t,\rho)$ and $\nu(t,\rho)$ are uniformly bounded for $0 \leqq t \leqq 1$ and for all large values of ρ. Substituting in (43), one has the following expression:

Case 1. $\beta \neq 0$,

*It is easily shown by a similar argument that there is a unique value $\mu_0(<\mu_1)$ such that when $\lambda = \mu_0$ the second boundary condition is satisfied.

$$u(z) = \beta \cos \rho z \, [1 - \frac{1}{2\rho} \int_0^z g(t) \sin 2\rho t \, dt + 0\,(\frac{1}{\rho^2})]$$

$$+ \frac{1}{\rho} \sin \rho z [-\alpha + \frac{\beta}{2} \int_0^z g(t)\, dt + \tfrac{1}{2} \int_0^z g(t) \cos 2\rho t \, dt$$

$$+ 0\,(\frac{1}{\rho})] \qquad (52)$$

where each of the terms denoted by $0\,(1/\rho n)$ has the meaning that it is bounded by <u>constant</u> for large values ρ and for all values of z in [0,1]. It should be remarked that the partial derivative of such a function with respect to z is:

$$0(\frac{1}{\rho^{n-1}}) \;.$$

<u>Case 2</u>. $\beta = 0$,

$$u(z) = -\alpha \sin \rho z \, [1 - \frac{1}{2\rho} \int_0^z g(t) \sin 2\rho t \, dt + 0(\frac{1}{\rho^2})]$$

$$+ \frac{\alpha}{\rho} \cos \rho z \, [\tfrac{1}{2} \int_0^z g(t)\, dt - \tfrac{1}{2} \int_0^z g(t) \cos 2\rho \, t dt$$

$$+ 0(\frac{1}{\rho^2})] \qquad (52a)$$

Let the (1+m)th eigenvalue by $\lambda_m = \frac{\rho_m^2}{k^2}$. For the case $\delta \neq 0$, the zeros of u(z), by the virtue of (52), are found from an equation of the form.

$$F(z,\rho) = \cot \rho z - 0(\frac{1}{\rho}) = 0 \qquad (53)$$

where $\frac{\partial}{\partial z} \; 0(\frac{1}{\rho}) = 0(1)$. It is then clear that for fixed large values of ρ, $\frac{\partial F(z,\rho)}{\partial z} < 0$. Thus, $F(z,\rho)$ is monotonically decreasing in z at those points where the function $F(z,\rho)$ is finite. Consequently, it has one zero in each interval of (ρz) of the length π. Since by the auxiliary theorem, this equation has exactly m zeros when

$\rho = \rho_m$, therefore:

$$\rho_m = m\pi + \varepsilon(m)\pi \tag{54}$$

where:

$$0 < \varepsilon(m) < 1 .$$

Now ρ_m is a root of (45), which, for the case $\delta\beta \neq 0$, can be rewritten in the form:

$$\tan \rho_m = \frac{c}{\rho_m} + 0(\frac{1}{\rho_m^2}) \tag{55}$$

where c is a constant. Thus:

$$\rho_m = \tan^{-1} \frac{c}{\rho_m} + 0(\frac{1}{\rho_m^2})$$

$$= m\pi + \frac{c}{\rho_m} + 0(\frac{1}{\rho_m^2})$$

$$= m\pi + \frac{c}{m\pi} + 0(\frac{1}{m^2}) \tag{56}$$

For the case $\delta = \beta = 0$, the asymptotic expression:

$$\rho_m = (m+1)\pi + \frac{c}{m\pi} + 0(\frac{1}{m^2})$$

For the case $\delta = 0$, $\beta \neq 0$

$$\rho_m = (m+\tfrac{1}{2})\pi + \frac{c}{m\pi} + 0(\frac{1}{m^2}) \tag{56a}$$

8.2 Proof of the Convergence of the Sturm-Liouville's Expansion

Let $\phi_m(z)$ be the eigenfunction corresponding to the eigenvalue ρ_m. From (52) or (52a) and (56) or (56a), the eigenfunction is seen to be of the form for all the cases:

$$\varphi_m(z) = B_m^{(1)}(z)\ [\sqrt{2} + \frac{h_1(z)}{m} + O(\frac{1}{m^2})] \quad (57)$$

$$+ \frac{1}{m} B_m^{(2)}(z)\ [c + h_2(z) + O(\frac{1}{m})$$

where (i) $B_m^{(1)}$ (z) and $B_m^{(2)}$ (z) are alternatively cos $p_m \pi z$ and sin $p_m \pi z$, (p_m = m, m + ½ or m + 1)

(ii) $h_j(z) = \int_0^z g(t)\ (c_j \cos p_m \pi t + d_i \sin p_m \pi t)\ dt$

+ a continuously differentiable function of z independent of m. (j = 1, 2); (58)

(iii) c, c_j, and d_i are constants independent of m. It is clear that the normalizing factor is:

$$\int_0^1 \varphi_m^2(z)\ dz = 1 + O(\frac{1}{m})$$

Now, let f(z) be any function of bounded variation* on [0,1]. Then f(z) can be expressed as a difference of two (bounded) positive, monotonically increasing function $\chi_1(z)$ and $\chi_2(z)$:

$$f(z) = \chi_1(z) - \chi_2(z) \quad (60)$$

Let the Fouriers coefficients of f(z) with respect to $B_m^{(1)}(z)$ and $B_m^{(2)}(z)$ be respectively $b_m^{(1)}$ and $b_m^{(2)}$. Now the generalized coefficient of f(z) with respect to the eigenfunction is:

$$A_m = \int_0^1 f(z)\ \varphi_m(z)\ dz$$

$$= \sqrt{2}\ b_m^{(1)} + c\frac{b_m^{(2)}}{2} + k_m = \sqrt{2}\ b_m^{(1)} + k_m' \quad (61)$$

---

*See, for instance, Titchmarsh, Theory of Functions, pp. 355-356. On the footnotes on p. 9 of Chapter I and p. 53 of Chapter II.

It is to show that $\sum_{m=1}^{\infty} k_m$ converges. Consider:

$$L_m = \int_0^1 \chi(z) B_m^{(i)}(z)\,dz \int_0^z g(t) B_m^{(j)}(t)\,dt$$

$$= \int_0^1 g(t) B_m^{(j)}(t)\,dt \int_t^1 \chi(x) B_m^{(i)}(z)\,dz\ .$$

By the second law of the mean for integrals, one concludes from the positive increasing monotonicity of the function $\chi(z)$:

$$\int_t^1 \chi(z) B_m^{(i)}(z)\,dz = \chi(1) \int_{t'}^1 B_m^{(i)}(z)\,dz$$

$$= O(\tfrac{1}{m}) \qquad (t < t' < 1)$$

Since $B_m^{(i)}(z)$ is either $\sin p_m \pi z$ or $\cos p_m \pi z$. Thus,

$$k_m = O(\frac{1}{m^2}) \tag{62}$$

Now a_m is the ordinary Fourier coefficient of a definite function of bounded variation and c is a constant. By Schwarz' inequality,

$$\sum_{m=1}^{\infty} \left|\frac{a_m}{m}\right|$$

converges. Thus from (61),

$$A_m = \sqrt{2}\ \ b_m^{(1)} + k_m' \tag{63}$$

where $\sum_{m=1}^{\infty} k_m'$ converges. Consequently, from (57)

$$A_m \varphi_m(z) = 2 b_m^{(1)} B_m^{(1)}(z) + R_m(z) \tag{64}$$

where $\sum_{m=1}^{\infty} R_m(z)$ converges uniformly in z on [0,1].

Therefore, if f(z) is continuous,

$$\sum_{m=1}^{\infty} b_m^{(1)} B_m^{(1)}(z)$$

converges uniformly in any closed sub-interval of [0,1].* By (59), the general Sturm-Liouville expansion has the same convergence properties when the eigenfunctions $\varphi_m(z)$ are normalized.

8.2.1 The closedness of the eigenfunctions. The convergence of the Sturm-Liouville expansion does not necessarily mean the limit of the expansion is $f(z)$ for $0 < z < 1$. In order to prove that

$$f(z) = \sum_{m=1}^{\infty} A_m \varphi_m(z) \qquad (0 < z < 1)$$

where $\varphi_m(z)$ are normalized and $f(z)$ is continuous and of bounded variation on [0,1], it is sufficient to show that the set of functions $\varphi_m(z)$ is closed.

Lemma.** If $\psi(x)$ is continuous on [0,1] and if:

$$\int_0^1 \psi(x)\varphi_n(x) \quad dx = 0$$

for all n, then $\psi(x) = 0$.

Proof. Consider the system

$$\frac{d^2 v}{dx^2} + \{\rho^2 - g(x)\} v + \psi(x) = 0 \tag{65}$$

with the boundary condition (40). When $\rho^2 \leqq$ the first eigenvalue ρ_1^2, the system has a unique solution which may be expressed in the form of the infinite series:

---

*Cf the theorem on p. 10, Chapter I and the exercise on p.107, Chapter III.

**This lemma can also be derived from the Hilbert theory of integral equation. The present elementary proof, due to Kneser, is based on the analytic property of the solution as a function the parameter = ρ^2/k^2.

$$v(x) = v_0 + \rho^2 v_1 + \rho^4 v_2 + \ldots + \rho^{2n} v_n + \ldots , \qquad (66)$$

where v_0, v_1 .... are function of x satisfying successively the differential equations:

$$\frac{v_0'' - g(x)v_0 + \psi(x) = 0 ,}{v_n'' - g(x)v_n + v_{n-1} = 0 .} \qquad (n = 1,2,.....) \qquad (66a)$$

Each of the functions satisfies the boundary conditions (40). From these equations and from the boundary conditions, one concludes by the integration in parts that:

$$\int_0^1 (v_m v_n - v_{m+1} v_{n-1})\, dx = \int_0^1 (v_{m+1} v_n'' - v_n v_{m+1}'')\, dx = 0$$

Put:

$$w_{m+1} = \int_0^1 v_m v_n\, dx = \int_0^1 v_{m+\nu} v_{n-\nu}\, dx$$

where $\nu \leqq m, n$. Now, for any real constants a and b,

$$0 \leqq \int_0^1 (av_{m-1} + bv_{m+1})^2\, dx = W_{2m-2} a^2 + W_{2m}\, ab + \bar{W}_{2m+2}\, b^2 .$$

Thus, the discriminant of the quadratic form in a and b must be negative; namely

$$W_{2m}^2 - W_{2m-2}\, W_{2m+2} \leqq 0 \qquad (67)$$

From the definition,

$$W_{2m} \geqq 0 .$$

Therefore, either all W_{2m} are zero or they are all positive. Suppose, now, $W_0 > 0$.

Then, from (67),

$$\frac{W_2}{W_0} \leqq \frac{W_4}{W_2} \leqq .... \leqq \frac{W_{2m}}{W_{2m-2}} \leqq .... \tag{68}$$

Now if the system (65) has a solution $v_n(x)$ for $\rho = \rho_n$, the nth eigenvalue, then

$$\int_0^1 \psi \cdot \varphi_n(x)\, dx = 0$$

as it is easily seen by multiplying both of the equations (39) and (65) alternatively the solutions and by taking the difference. The converse is also true.*

By the assumption that

$$\int_0^1 \psi(x)\varphi_n(x)\, dx = 0$$

for all n, the system has a solution for all values of ρ. By the fundamental existence theorem this solution is analytic for all values of ρ. Thus, the series (66) would converge for all values of ρ and all x on [0,1]. Consequently,

$$\int_0^1 v_0\, v\, dx = W_0 + \rho^2 W_1 + \rho^4 W_2 + ...$$

is finite for all values of ρ. Yet, it is impossible in view of (68). Thus, it follows that:

$$W_0 = W_2 = ... = W_{2n} = ... = 0 .$$

Consequently, $v_0 = 0$ and therefore from (66a), $\psi(x) = 0$.

Now consider:

$$\psi(x) = f(x) - \sum_{n=1}^{\infty} A_n \varphi_n(x) \quad \text{for} \quad 0 < x < 1$$

---

*This can be proved by a similar reasoning as in Section 8. See, also for instance, Ince, pp. 266-269.

and define:

$$\psi(0) = \psi(0+) \quad \text{and} \quad \psi(1) = \psi(1-0) .$$

The latter limits exist because $f(x)$ in the preceding section is of bounded variation. $\psi(x)$ is continuous. By the exercise (1) on p. 14 of Chapter I, the series is boundedly convergent, it can be integrated term by term. Therefore:

$$\int_0^1 \psi(x)\, \varphi_n(x)\, dx = 0 \quad \text{for all } n$$

Hence:

$$\psi(x) = 0 \quad \text{for} \quad 0 < x < 1 .$$

This completes the Sturm-Liouville expansion theorem.

REFERENCES

I. Classical Texts on Mathematical Analysis

(1) Goursat-Hedrick, Mathematical Analysis, Vol. 1, 1917.

(2) E.C. Titchmarsh, Theory of Functions, Oxford U.P., New York, 1950.

(3) E.T. Whittaker and G. Watson, A Course on Modern Analysis, Cambridge U.P., Cambridge, 1952.

II. Ordinary Differential Equations and Applications of Fourier Series to Boundary-Value Problems

(1) E.L. Ince, Ordinary Differential Equations, Longmans, Green & Co., London, New York, 1927.

(2) H.S. Carslaw and J.C. Jaeger, Conduction of Heat in a Solid, Oxford U.P., Oxford, 1959.

(3) R.V. Churchill, Fourier Series and Boundary Value Problems, McGraw-Hill, New York, 1963.

III. Eigen-Value Problems, Orthogonal Functions, and Special Functions

(1) L. Collatz, Eigenwertproblem und ihre Numerische behandlung, Academische Verlags Gesellschaft, 1945.

(2) D. Jackson, Fourier Series and Orthogonal Functions, Mathematical Association of American Monograph, 1941.

(3) E. Artin, The Gamma Function, Holt, Rinehart and Winston, New York, 1964.

(4) W. Magnus, F. Oberhettinger, and R.P. Soni, Formulas and Theorems for the Special Functions of Mathematical Physics, Springer-Verlag, New York, 1966.

IV. Methods of Mathematical Physics and Partial Differential Equations

(1) H. Bateman, Partial Differential Equations of Mathematical Physics, Cambridge U.P., Cambridge, 1959.

(2) R. Courant and D. Hilbert, Methods of Mathematical Physics, Vol. 1, Interscience, New York, 1953.

(3) J. Jeffreys and B. Jeffreys, Method of Mathematical Physics, Cambridge U.P., Cambridge, 1956.

(4) P.M. Morse and H. Feshbach, Methods of Theoretical Physics, Vols. 1 and 2, McGraw-Hill, 1953.

(5) A.J.D. Sommerfeld, Partial Differential Equations in Physics, Academic, New York, 1953.

(6) I.S. Sokolnikoff and R.M. Redheffer, Mathematics of Physics and Engineering, McGraw-Hill, New York, 1966.

V. Fourier Transforms and Operational Calculus

(1) I.N. Sneddon, Fourier Transforms, McGraw-Hill, 1951.

(2) E.C. Titchmarsh, Theory of Fourier Integrals, Oxford U.P., New York, 1937.

(3) H.S. Carslaw and J.C. Jaeger, Operational Methods in Applied Mathematics, Oxford U.P., New York, 1948.

(4) N.W. McLachlan, Modern Operational Calculus, MacMillan, London, 1948.

(5) A. Erdelyi, W. Magnus, F. Oberhettinger and F.H. Tricomi, Tables of Integral Transforms, (2 Vols.), McGraw-Hill, New York, 1954.

VI. Potential Theory and Residue Theory

(1) O.D. Kellogg, Foundations of Potential Theory, Berlin Springer, 1929.

(2) E.T. Copson, An Introduction to the Theory of Functions of a Complex Variable, Oxford, The Clarendon Press, 1935.